Lecture Notes in Computational Vision and Biomechanics

Volume 33

Series Editors

João Manuel R. S. Tavares⦾, Departamento de Engenharia Mecânica, Faculdade de Engenharia, Universidade do Porto, Porto, Portugal
Renato Natal Jorge, Faculdade de Engenharia, Universidade do Porto, Porto, Portugal

Research related to the analysis of living structures (Biomechanics) has been carried out extensively in several distinct areas of science, such as, for example, mathematics, mechanical, physics, informatics, medicine and sports. However, for its successful achievement, numerous research topics should be considered, such as image processing and analysis, geometric and numerical modelling, biomechanics, experimental analysis, mechanobiology and Enhanced visualization, and their application on real cases must be developed and more investigation is needed. Additionally, enhanced hardware solutions and less invasive devices are demanded.

On the other hand, Image Analysis (Computational Vision) aims to extract a high level of information from static images or dynamical image sequences. An example of applications involving Image Analysis can be found in the study of the motion of structures from image sequences, shape reconstruction from images and medical diagnosis. As a multidisciplinary area, Computational Vision considers techniques and methods from other disciplines, like from Artificial Intelligence, Signal Processing, mathematics, physics and informatics. Despite the work that has been done in this area, more robust and efficient methods of Computational Imaging are still demanded in many application domains, such as in medicine, and their validation in real scenarios needs to be examined urgently.

Recently, these two branches of science have been increasingly seen as being strongly connected and related, but no book series or journal has contemplated this increasingly strong association. Hence, the main goal of this book series in Computational Vision and Biomechanics (LNCV&B) consists in the provision of a comprehensive forum for discussion on the current state-of-the-art in these fields by emphasizing their connection. The book series covers (but is not limited to):

- Applications of Computational Vision and Biomechanics
- Biometrics and Biomedical Pattern Analysis
- Cellular Imaging and Cellular Mechanics
- Clinical Biomechanics
- Computational Bioimaging and Visualization
- Computational Biology in Biomedical Imaging
- Development of Biomechanical Devices
- Device and Technique Development for Biomedical Imaging
- Experimental Biomechanics
- Gait & Posture Mechanics
- Grid and High Performance Computing on Computational Vision and Biomechanics
- Image Processing and Analysis
- Image processing and visualization in Biofluids
- Image Understanding
- Material Models
- Mechanobiology
- Medical Image Analysis

- Molecular Mechanics
- Multi-modal Image Systems
- Multiscale Biosensors in Biomedical Imaging
- Multiscale Devices and BioMEMS for Biomedical Imaging
- Musculoskeletal Biomechanics
- Multiscale Analysis in Biomechanics
- Neuromuscular Biomechanics
- Numerical Methods for Living Tissues
- Numerical Simulation
- Software Development on Computational Vision and Biomechanics
- Sport Biomechanics
- Virtual Reality in Biomechanics
- Vision Systems
- Image-based Geometric Modeling and Mesh Generation
- Digital Geometry Algorithms for Computational Vision and Visualization

In order to match the scope of the Book Series, each book has to include contents relating, or combining both Image Analysis and mechanics. Indexed in SCOPUS, Google Scholar and SpringerLink.

More information about this series at http://www.springer.com/series/8910

João Manuel R. S. Tavares ·
Paulo Rui Fernandes

Editors

New Developments
on Computational Methods
and Imaging in Biomechanics
and Biomedical Engineering

 Springer

Editors
João Manuel R. S. Tavares **ⒾⒹ**
Departamento de Engenharia Mecânica
Faculdade de Engenharia
Universidade do Porto
Porto, Portugal

Paulo Rui Fernandes **ⒾⒹ**
Department of Mechanical Engineering
IDMEC-IST
Lisbon, Portugal

ISSN 2212-9391   ISSN 2212-9413   (electronic)
Lecture Notes in Computational Vision and Biomechanics
ISBN 978-3-030-23072-2   ISBN 978-3-030-23073-9   (eBook)
https://doi.org/10.1007/978-3-030-23073-9

This Springer imprint is published by the registered company Springer Nature Switzerland AG
The registered company address is: Gewerbestrasse 11, 6330 Cham, Switzerland

Preface

The 15th International Symposium on Computer Methods in Biomechanics and Biomedical Engineering and the 3rd Conference on Imaging and Visualization (CMBBE2018), were run together at Instituto Superior Técnico (IST), Technical University of Lisbon, Portugal, from March 20 to 26, 2018.

We believe that CMBBE2018 had a strong impact on the development of computational biomechanics and biomedical imaging and visualization; particularly, by identifying emerging areas of research and promoting the collaboration and networking between participants. Actually, CMBBE2018 included 176 oral presentations and 37 poster presentations. In addition, 16 renowned researchers delivered very interesting plenary keynotes, addressing current challenges in computational biomechanics and biomedical imaging. This book includes the extended versions of selected works presented in CMBBE2018.

Briefly, the included 10 chapters address important topics in Biomechanics and Biomedical Imaging, including Control Theory, Finite Element Method, Fluid Dynamics, Geometrical Modeling, Image Segmentation, Image Analysis, Monte Carlo Simulation, Multibody Modeling, and Numerical Methods. Different applications are addressed and described throughout the book, comprising Computational Simulation, Flow Analysis, Medical Diagnosis and Rehabilitation, Numeral Analysis, and Stress and Strain Analysis.

Therefore, this book is of high interest for Researchers, Students, End Users, and Manufacturers from several multidisciplinary fields, as the ones related with Bioengineering, Biology, Biomechanics, Computational Mechanics, Computer Graphics, Computer Sciences, Mathematics, Mechanobiology, Medical Imaging, Medicine, Physics, Physiological Cybernetics, and Telemetry.

The editors would like to take this opportunity to thank the authors of the 10 selected contributions for sharing their works, experiences, and knowledge, making possible their dissemination through this book.

Porto, Portugal   João Manuel R. S. Tavares
Lisbon, Portugal   Paulo Rui Fernandes
                                                        Co-editors and Co-chairs of CMBBE2018

The original version of this book was revised: Volume number has been corrected. The correction to this book can be found at https://doi.org/10.1007/978-3-030-23073-9_11

Contents

About the Editors

João Manuel R. S. Tavares Departamento de Engenharia Mecânica, Faculdade de Engenharia Universidade do Porto, Porto, Portugal
Email: tavares@fe.up.pt
url: www.fe.up.pt/∼tavares

He graduated in Mechanical Engineering at the Universidade do Porto, Portugal in 1992. He also earned his M.Sc. degree and Ph.D. degree in Electrical and Computer Engineering from the Universidade do Porto in 1995 and 2001, and attained his Habilitation in Mechanical Engineering in 2015. He is a senior researcher at the Instituto de Ciência e Inovação em Engenharia Mecânica e Engenharia Industrial (INEGI) and Associate Professor with Habilitation at the Department of Mechanical Engineering (DEMec) of the Faculdade de Engenharia da Universidade do Porto (FEUP).

João Tavares is co-editor of more than 45 books, co-author of more than 35 book chapters, 650 articles in international and national journals and conferences, and 3 international and 3 national patents. He has been a committee member of several international and national journals and conferences, is co-founder and co-editor of the book series "Lecture Notes in Computational Vision and Biomechanics" published by Springer, founder and Editor-in-Chief of the journal "Computer Methods in Biomechanics and Biomedical Engineering: Imaging & Visualization" published by Taylor & Francis, and co-founder and co-chair of the international conference series: CompIMAGE, ECCOMAS VipIMAGE, ICCEBS, and BioDental. Additionally, he has been (co-)supervisor of several M.Sc. and Ph.D. theses and supervisor of several postdoc projects, and has participated in many scientific projects both as researcher and as scientific coordinator.

His main research areas include computational vision, medical imaging, computational mechanics, scientific visualization, human–computer interaction, and new product development (more information can be found at www.fe.up.pt/∼tavares)

Paulo Rui Fernandes Instituto Superior Técnico, Universidade de Lisboa, Lisbon, Portugal

IDMEC, Instituto de Engenharia Mecânica, Lisboa, Portugal

Email: paulo.rui.fernandes@tecnico.ulisboa.pt

url: https://fenix.tecnico.ulisboa.pt/homepage/ist13157

He is Associate Professor with "Agregação" (Habilitation) at Mechanical Eng. Department of Instituto Superior Técnico (IST), University of Lisbon, where he received his Ph.D. degree in 1998 and the "Agregação" in 2012. After the Ph.D., he was a postdoctoral fellow at the Musculoskeletal Research Laboratory of Pennsylvania State University from July to December 1998. He was awarded with the IBM scientific prize 1997 (IBM— Portugal) with the work "Simulation of the Bone Remodelling Process". His main research field is Biomechanics, particularly Bone Tissue Mechanics and Orthopaedic Implants Design. Paulo Fernandes teaches Computational Mechanics and Tissue Biomechanics in the Biomedical Engineering Program where he was the coordinator of the Biomechanics and Biomedical Devices profile from 2013 to 2018. His research work has been an important contribution for the development of Biomechanics in IST leading a research group with a strong impact in the formation of human resources. He has also led several research projects funded by the Portuguese Foundation for Science and Technology. He is author/co-author of numerous publications in international and national journal, book chapters, and communications in international conferences. He maintains international collaborations with research groups from Europe, Brazil, and the USA. He was a member of the council of the European Society of Biomechanics since July 2012–July 2016 and President of the Portuguese Society of Biomechanics from February 2013 to February 2017.

The Impact of Patches on Blood Flow Disorders In Carotid Artery

M. Ciałkowski, N. Lewandowska, M. Micker, M. Warot, A. Frąckowiak and P. Chęciński

Abstract The atherosclerotic plaques are surgically removed by endarterectomy of the common and internal carotid artery wall, removal of lesions, and suturing the artery again. To avoid arterial lumen stenosis, sewing a patch in the incision area is indicated, which will cause a slight expansion of the flow lumen. The channel expansion causes a positive tension gradient, enhancing separation of the parietal layer and occurrence of whirlpools. The latter may cause plaque redeposition. The selection of the patch size is not described in detail in the literature and is based on the surgeon's experience and intuition. The purpose of the studies is to determine the maximum patch width per surgical incision at which no flow separation will occur. To determine the geometry of the channel with a patch sewn in, an equation was determined to reflect the course of the arterial wall curves by math functions. The artery radius, the maximum expansion radius, and the length of the patch sewn in have been assumed as the input parameters that define the boundary conditions necessary for the determination of polynomial coefficients. By a gradual increase of the maximum radius, a geometry group was determined, which was the starting point for numerical simulations. The simulations were made with the use of Fluent. The increasing of the maximum radius was continued until the separation of the parietal layer was detected and whirlpools occurred. The results showed that when the maximum radius is 30% greater in relation to the arterial radius, whirlpools occur, which in consequence may lead to plaque redeposition.

M. Ciałkowski · N. Lewandowska (✉)
Faculty of Machines and Transport, Poznan University of Technology, Chair of Thermal Engineering, Piotrowo 3, 60-695 Poznan, Poland
e-mail: natalia.lewandowska.pp@gmail.com

M. Micker · M. Warot · A. Frąckowiak · P. Chęciński
Department of General and Vascular Surgery and Angiology, Poznan University of Medical Sciences (PUMS), 34 Dojazd St, 60-631 Poznan, Poland

© Springer Nature Switzerland AG 2019
J. M. R. S. Tavares and P. R. Fernandes (eds.), *New Developments on Computational Methods and Imaging in Biomechanics and Biomedical Engineering*, Lecture Notes in Computational Vision and Biomechanics 33, https://doi.org/10.1007/978-3-030-23073-9_1

1 Introduction

Atherosclerotic plaques very often deposit at the points of carotid artery bifurcation, which might result in a closure of the arterial lumen. According to the experimental research [1, 2, 3], when the stenosis level in the arterial lumen reaches ca. 60–75%, it is necessary to remove the arterial lesions surgically. This operation is a generally accepted form of primary or secondary ischemic stroke prophylactics [4]. The plaques are most frequently removed by means of lengthwise incision of the artery wall, removal of the adhering plaque and resuturing of the artery. It is also possible to cut off the internal carotid artery, extract of the atherosclerotic plaque and reconnect the cutoff artery. This procedure, however, is less frequent. Direct suturing of the membrane (Fig. 1a) causes artery lumen stenosis. This is a negative phenomenon as it reduces the arterial flow capacity and may cause disorders. Both factors mentioned above cause an increased restenosis probability. Considering the said prerequisites, experienced surgeons recommend insertion at the point of the artery incision of a patch (Fig. 1b) made of plastic (usually dacron or polytetrafluoroethylene—PTFE) or tissue taken from the patient (most frequently a vein). Dacron, however, is the most commonly used material. The application of a patch reduces the risk of a stroke, death, or restenosis as compared to the primary suturing of the wound after the arteriotomy [5, 6]. It eliminates the risk of arterial lumen stenosis, but in turn, causes its expansion. The patch width is selected directly on the operating table when the surgeon adjusts the appropriate patch geometry based on his/her own experience. There is no scientific justification of the selected patch geometry—its width is fixed intuitively in most cases. This paper presents an attempt of an analytic and mathematical determination of the geometry of the patches that minimize the risk of restenosis.

From the mechanical point of view, the flow through the divergent canal enhances the separation of the layer adhering to the wall and a formation of a reverse flow near the walls. The occurring whirlpools may cause 'suction' of solid particles into

(a)   **(b)**

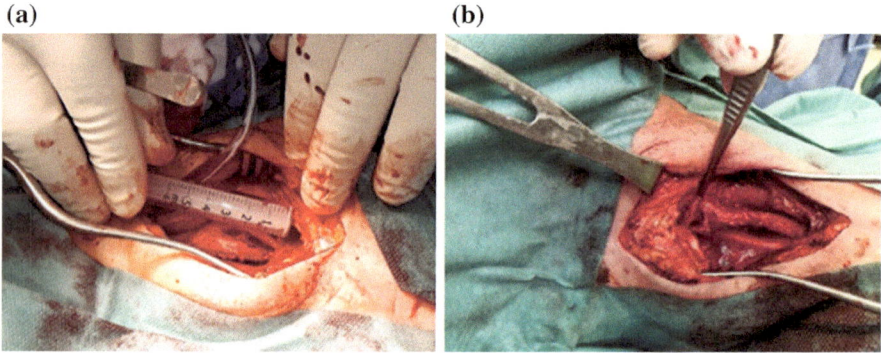

Fig. 1 Plague removal surgery: **a** with a patch **b** without a patch

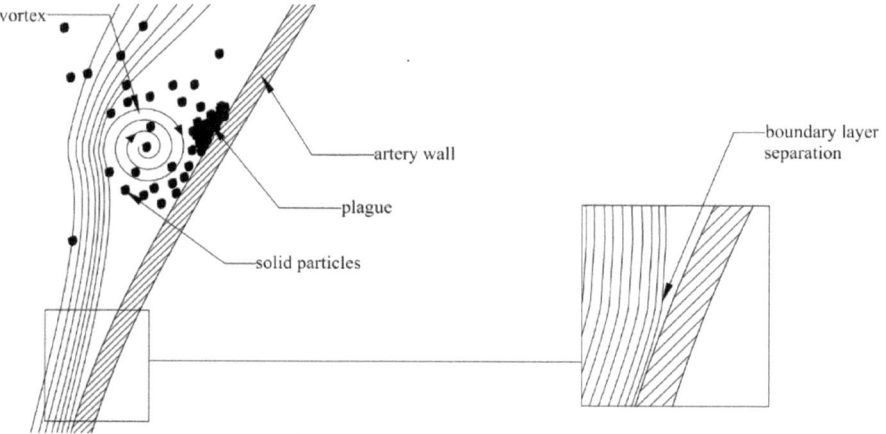

Fig. 2 Schematics of the basic concept of boundary layer separation

the center, which in consequence, may lead to plaque redeposition near the wall. Figure 2 presents the concept of this phenomenon.

Upon respective expansion of the arterial canal, whirlpools occur in the blood flow, which in consequence may lead to plaque redeposition on the walls.

The purpose of the studies is to determine the geometry of a patch that would not cause flow turbulization when inserted in the carotid artery. The research works considered analytical and numerical computation. The developed geometric models were based on the analytical results and represented a basis for the numerical computation. A series of simulations was carried out to show the impact of the geometry on the parameters characterizing the flow. The analysis results showed that whirlpools begin to appear near the walls when the artery diameter is expanded by more than 30%, compared to its original dimension. This corresponds to the maximum patch width of 10–14 mm, depending on the arterial diameter. The geometry of the patch that would not lead to restenosis has been developed based on the obtained results. The research results represent the first fully documented analysis concerning the geometry of the inserted patches. They may constitute the foundations for the selection of a specific patch width, customized to the geometry of a patient's artery.

2 Methods

2.1 Mathematical Analysis

The patch geometry was determined by means of three functions: fourth-degree polynomial, a spline function composed of two cubic polynomials and the ellipsis

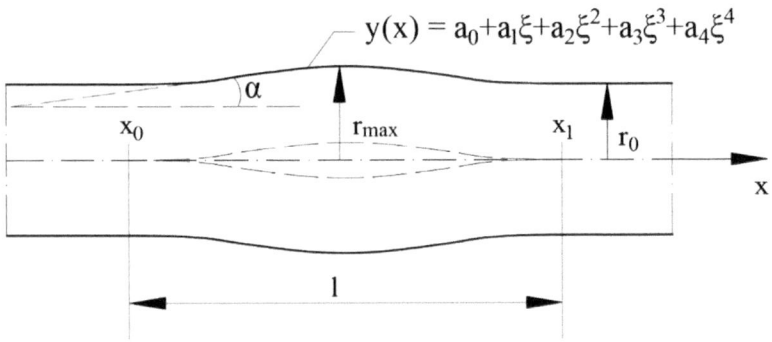

Fig. 3 Patch geometry

function. The numerical tests showed that the ellipsis is the least favorable function. Both polynomial functions under analysis yielded similar results.

2.1.1 Fourth-Degree Polynomial

The following fourth-degree polynomial was used to describe the expanded artery (Fig. 3):

$$y(x) = a_0 + a_1 \frac{x - x_0}{x_1 - x_0} + a_2 \left(\frac{x - x_0}{x_1 - x_0} \right)^2 + a_3 \left(\frac{x - x_0}{x_1 - x_0} \right)^3 + a_4 \left(\frac{x - x_0}{x_1 - x_0} \right)^4 \quad (1)$$

Introduction of a dimensionless variable $\xi = (x - x_0)/(x_1 - x_0)$ results in the following function:

$$y(\xi) = a_0 + a_1 \xi + a_2 \xi^2 + a_3 \xi^3 + a_4 \xi^4 \quad (2)$$

To determine the polynomial coefficients, the following boundary conditions have been applied:

- smooth connection of the part of unvaried diameter with the expanded part:

$$y(\xi = 0) = y(\xi = 1) = r_0 \quad (3)$$

- equality of tangents:

$$y'(\xi = 0) = y'(\xi = 1) = 0 \quad (4)$$

- maximum artery expansion radius:

$$y\left(\xi = \frac{1}{2}\right) = r_{max} \tag{5}$$

After implementing above boundary conditions, polynomial coefficients values are equal:

$$\begin{cases} a_0 = r_0 \\ a_1 = 0 \\ a_2 = 16(r_{max} - r_0) \\ a_3 = -32(r_{max} - r_0) \\ a_4 = 16(r_{max} - r_0) \end{cases} \tag{6}$$

Upon determination of the polynomial coefficients, function assumes the following form:

$$y(\xi) = r_0 + 16 \cdot (r_{max} - r_0) \cdot \xi^2 \cdot (1 - \xi)^2, \xi = \frac{x - x_0}{x_1 - x_0} = \frac{x - x_0}{1}, x = x_0 + 1 \cdot \xi \tag{7}$$

The angle of inclination of the variable edge to the x-axis is

$$tg\alpha(\xi) = \frac{dy}{dx} = \frac{dy}{d\xi} \cdot \frac{d\xi}{dx} = \frac{dy}{1d\xi} = \frac{1}{1} \cdot 32(r_{max} - r_0)\xi(1 - \xi) \cdot (1 - 2\xi) \tag{8}$$

while the extreme inclination angle results from zeroing of I derivative, namely

$$(tg\alpha(\xi))' = 32(r_{max} - r_0)\left(1 - 6\xi + 6\xi^2\right)/1 = 0 \tag{9}$$

which occurs for

$$\xi_1^* = \frac{1}{2} - \frac{1}{2} \cdot \frac{1}{\sqrt{3}} = \frac{1}{2}\left(1 - \frac{1}{\sqrt{3}}\right) \approx 0.211 \tag{10}$$

$$\xi_2^* = \frac{1}{2}\left(1 + \frac{1}{\sqrt{3}}\right) \approx 0.789 \tag{11}$$

whereas

$$\left(tg\alpha\left(\xi_1^*\right)\right)'' = -192(r_{max} - r_0) \cdot \left(1 - 2\xi_1^*\right)/1 < 0 \tag{12}$$

therefore angle $\alpha\left(\xi_1^*\right)$ reaches its maximum value.

For $\xi = \xi_2^*$, the tangent function drops to its minimum value, because angle α is an obtuse angle.

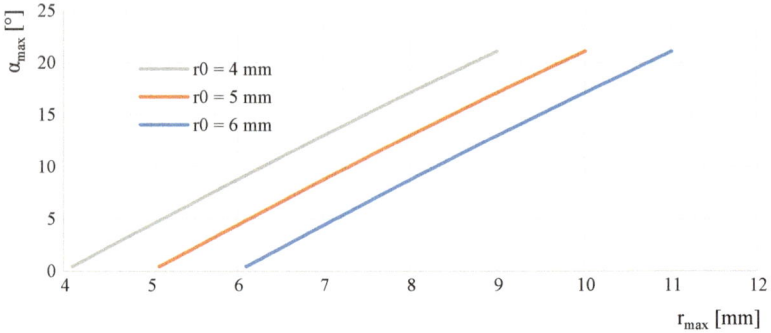

Fig. 4 Function of the maximum angle of inclination for different value of artery radius (r_0) and the constant value of patch width l = 40 mm

Parameter r/r_{max} is related to angle $\alpha_{max} = \alpha\left(\xi = \xi_1^*\right) = \alpha\left(\xi = \frac{1}{2}\left(1 - \frac{1}{\sqrt{3}}\right)\right)$

$$tg\alpha_{max} = 32\frac{r_{max} - r_0}{l} \cdot \xi(1 - \xi)(1 - 2\xi) \tag{13}$$

$$tg\alpha_{max} = 32\frac{r_{max} - r_0}{l}\frac{1}{2}\left(1 - \frac{1}{\sqrt{3}}\right)\frac{1}{2}\left(1 + \frac{1}{\sqrt{3}}\right)\frac{1}{\sqrt{3}} = \frac{16}{3\sqrt{3}} \cdot \frac{r_{max} - r_0}{l} \tag{14}$$

$$\alpha_{max} = \text{arc tg}\frac{16}{3\sqrt{3}} \cdot \frac{r_{max} - r_0}{l} = \text{arc tg}\frac{16}{3\sqrt{3}} \cdot \frac{r_0}{l}\left(\frac{r_{max}}{r_0} - 1\right) \tag{15}$$

or

$$r_{max} = r_0 + \frac{3\sqrt{3}}{16} \cdot l \cdot tg\alpha_{max} \tag{16}$$

Angle α_{max} for the outline described by function (Eq. 15) will result from a flow without separation, being the Reynolds number function. As initial value for further calculation there will be assumed $\alpha_{max} = 10°$ then

$$r_{max} = r_0 + 3\sqrt{3} \cdot 16^{-1} \cdot 0.003046 \cdot l = r_0 + 0.00099 \cdot l \tag{17}$$

for $\alpha_{max} = 45°$ l = 40 mm, $r_0 = 5$ mm

$$r_{max} = r_0 + \frac{3\sqrt{3}}{16} \cdot l \cdot 1 = r_0 + 0.3248 \, l$$

$$r_{max} = 5 + 13 = 18 \text{ mm}, \quad r_{max}/r_0 = 1 + 2.6 = 3.6,$$

which is an unreal result. In Figs. 4, 5, 6, 7 charts $\alpha_{max} = f(r_{max})$ and $r_{max} = f(r_0, \alpha_{max})$ are presented.

Patch width

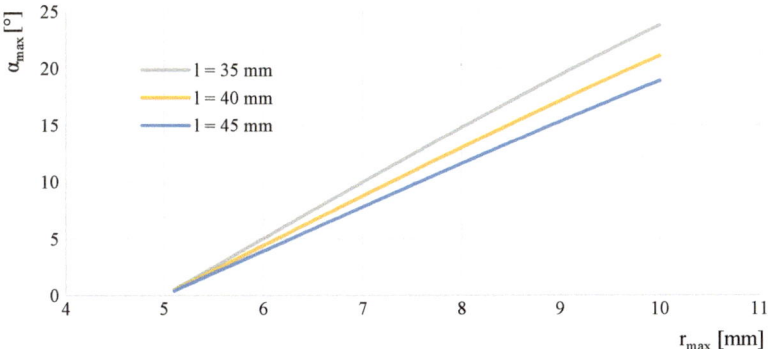

Fig. 5 Function of the maximum angle of inclination for different value of patch length (l) and the unchanged value of artery diameter ($r_0 = 5$ mm)

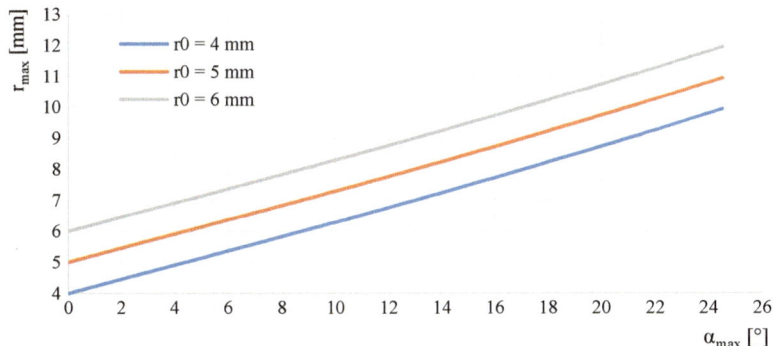

Fig. 6 Function of the maximum radius in the widening part of artery for different values of artery radius (r_0) and the patch length $l = 40$ mm

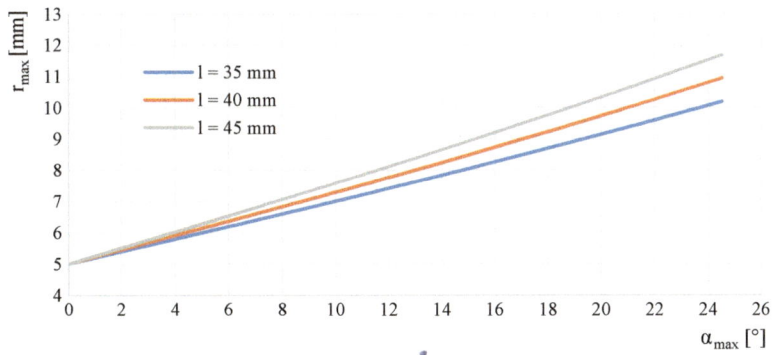

Fig. 7 Function of the maximum radius in the widening part of artery for different values of patch length (l) and the artery radius $r_0 = 5$ mm

The perimeter patch width is the result of the relation

$$s(\xi) = 2\pi \cdot y(\xi) - 2\pi r_0 = 32\pi(r_{max} - r_0)\xi^2(1 - \xi)^2 \tag{18}$$

and reaches its maximum in the center $\xi = 0.5$

$$s_{max} = 2\pi(r_{max} - r_0) \tag{19}$$

or

$$\frac{s_{max}}{2\pi r_0} = \frac{r_{max}}{r_0} - 1 \tag{20}$$

The parameter $\zeta = r_{max}/r_0$ was introduced to the numerical computation, then

$$y(\xi) = r_0 \cdot \left[1 + 16\left(\frac{r_{max}}{r_0} - 1\right)\xi^2(1 - \xi)^2\right] = r_0 \cdot \left[1 + 16(\zeta - 1)\xi^2(1 - \xi)^2\right] \tag{21}$$

or in the dimensionless form

$$\frac{y(\xi)}{r_0} = 1 + 16(\zeta - 1)\xi^2(1 - \xi)^2 \tag{22}$$

$$\alpha_{max} = arc\, tg\left(\frac{16}{3\sqrt{3}} \cdot \frac{r_0}{1}(\zeta - 1)\right) \tag{23}$$

$$\frac{s_{max}}{2\pi} = r_0(\zeta - 1) \rightarrow \frac{s_{max}}{2\pi r_0} = \zeta - 1 \tag{24}$$

value $s_{max}/2\pi$ expresses the artery radius increase at the widest point of the patch insertion, while $s_{max}/2\pi r_0$ is the dimensionless growth of the radius.

Assuming that $r_0 = 5$ mm, $l = 40$ mm, $\zeta \leq 1.3$ the argument of function (21) is

$$\frac{16}{3\sqrt{3}} \cdot \frac{5}{40} \cdot (1.3 - 1) = 0.11547 \Rightarrow arc\, tg\, 0.11547 = 0.1149\ rd = 6.587°,$$

on the other hand

$$arc\, tg\, x = x - \frac{x^3}{3} + \frac{x^5}{5} = 0.11547 - 0.00051 = 0.11493\ rd$$

while the omission of higher exponents leads to expression

$$arc\, tg\, x = x \cdots = 0.11547\ rd = 6.616°$$

which produces the relative error

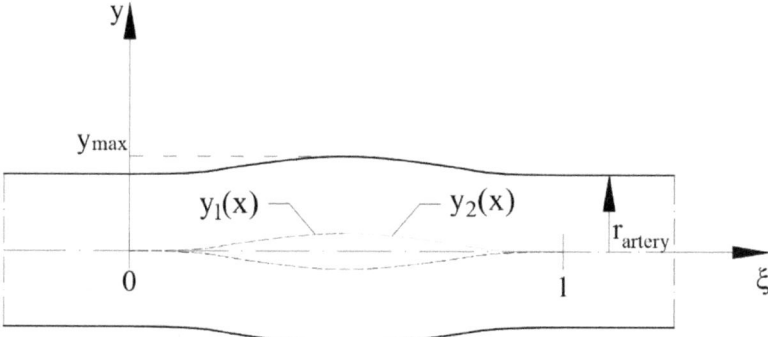

Fig. 8 Analytical description of the spline function

$$\delta = \frac{6.587° - 6.616°}{6.587°} \cdot 100\% \approx 0.44\%$$

So, in the interval $\alpha_{max} \leq 6,5°$, we obtain a simple dependency between the maximum radius and the corresponding maximum inclination angle of the tangent to the artery outline

$$\alpha_{max} = \frac{16}{3\sqrt{3}} \cdot \frac{1}{8}\left(\frac{r_{max}}{5} - 1\right) \rightarrow r_{max} = 5\left(1 + \frac{3\sqrt{3}}{2}\alpha_{max}\right) \qquad (25)$$

The patch width in the widest part of the artery (for $r_0 = 5$ mm and $\zeta = 1.3$)

$$s_{max} = 2\pi r_0(\zeta - 1) = 2\pi \cdot 5(1, 3 - 1) = 9.42 \text{ mm}$$

2.1.2 Third-Degree Polynomial Spline

Patch, in this case, is described by two third-degree polynomial functions (Fig. 8):

$$y_1(\xi) = a_0 + a_1\xi + a_2\xi^2 + a_3\xi^3 \qquad (26)$$

$$y_2(\xi) = b_0 + b_1\xi + b_2\xi^2 + b_3\xi^3 \qquad (27)$$

Conditions on the curve ends

$$y_1(0) = a_0 = 0 \qquad (28)$$

$$y_1'(0) = a_1 = 0 \qquad (29)$$

$$y_2(1) = b_0 + b_1 + b_2 + b_3 = 0 \qquad (30)$$

$$y_2'(1) = b_1 + 2b_2 + 3b_3 = 0 \tag{31}$$

$$y_1\left(\xi = \frac{1}{2}\right) = y_2\left(\xi = \frac{1}{2}\right) = r_{max} \tag{32}$$

which leads to

$$y_1(\xi) = a_2\xi^2 + a_3\xi^3 \tag{33}$$

$$y_1'(\xi) = 2a_2\xi + 3a_3\xi^2 \tag{34}$$

Convergence conditions at center point $\xi = 1/2$:

$$y_1\left(\frac{1}{2}\right) = \frac{1}{4}a_2 + \frac{1}{8}a_3 = y_{max} \tag{35}$$

$$y_1\left(\frac{1}{2}\right) = y_2\left(\frac{1}{2}\right) \Rightarrow \frac{1}{4}a_2 + \frac{1}{8}a_3 = b_0 + \frac{1}{2}b_1 + \frac{1}{4}b_2 + \frac{1}{8}b_3 \tag{36}$$

$$y_1'\left(\frac{1}{2}\right) = y_2'\left(\frac{1}{2}\right) \Rightarrow a_2 + \frac{3}{4}a_3 = b_1 + b_2 + \frac{3}{4}b_3 \tag{37}$$

$$y_1''\left(\frac{1}{2}\right) = y_2''\left(\frac{1}{2}\right) \Rightarrow 2a_2 + 6a_3 \cdot \frac{1}{2} = 2b_2 + 3b_3 \tag{38}$$

We have 8 equations with 8 unknowns ($a_0 = a_1 = 0$).

$$\begin{matrix} b_0 + b_1 + b_2 + b_3 = 0 \\ b_1 + 2b_2 + 3b_3 = 0 \\ \frac{1}{4}a_2 + \frac{1}{8}a_3 = y_{max} \\ \frac{1}{4}a_2 + \frac{1}{8}a_3 = b_0 + \frac{1}{2}b_1 + \frac{1}{4}b_2 + \frac{1}{8}b_3 \\ a_2 + \frac{3}{4}a_3 = b_1 + b_2 + \frac{3}{4}b_3 \\ 2a_2 + 3a_3 = 2b_2 + 3b_3 \end{matrix} \tag{39}$$

$$\begin{bmatrix} \frac{1}{4} & \frac{1}{8} & 0 & 0 & 0 & 0 \\ \frac{1}{4} & \frac{1}{8} & -1 & -\frac{1}{2} & -\frac{1}{4} & -\frac{1}{8} \\ 1 & \frac{3}{4} & 0 & -1 & -1 & -\frac{3}{4} \\ 2 & 3 & 0 & 0 & -2 & -3 \\ 0 & 0 & 1 & 1 & 1 & 1 \\ 0 & 0 & 0 & 1 & 2 & 3 \end{bmatrix} \begin{Bmatrix} a_2 \\ a_3 \\ b_0 \\ b_1 \\ b_2 \\ b_3 \end{Bmatrix} = \begin{bmatrix} 1 & 0 & 0 & 0 & 0 & 0 \\ 0 & & & & & \vdots \\ 0 & & & & & \vdots \\ 0 & & & & & \vdots \\ 0 & & & & & \vdots \\ 0 & \cdots & \cdots & \cdots & \cdots & 0 \end{bmatrix} \begin{Bmatrix} y_{max} \\ 0 \\ \vdots \\ \vdots \\ \vdots \\ 0 \end{Bmatrix}$$

The solution of the system of equations is as follows:

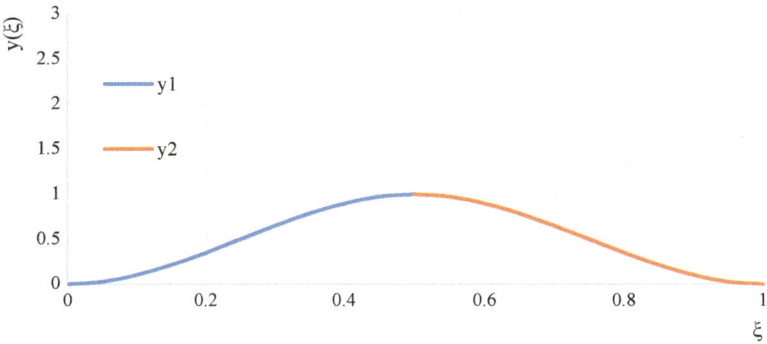

Fig. 9 Function of patch geometry composed of two third-polynomial functions described by Eqs. 41, 42

$$a_0 = a_1 = 0 \quad a_2 = 12 \cdot y_{max} \quad a_3 = -16 \cdot y_{max}$$
$$b_0 = -4 \cdot y_{max} \quad b_1 = 24 \cdot y_{max} \quad b_2 = -36 \cdot y_{max} \quad b_3 = 16 \cdot y_{max} \tag{40}$$

Third-degree curves that fulfil the conditions of zeroing the I derivative in $\xi = 0$ and $\xi = 1$ are described by functions:

$$y_1(\xi) = 12y_{max}\xi^2 - 16y_{max}\xi^3 = 4y_{max}\xi^2(3 - 4\xi) \tag{41}$$

$$y_2(\xi) = -4y_{max} + 24y_{max}\xi - 36y_{max}\xi^2 + 16y_{max}\xi^3 \tag{42}$$

In Fig. 9, the curve composed of Eqs. 41, 42 and for $y_{max} = 1$ is presented.

The maximum curve y_1 inclination angle is at the point where the first derivative of the tangent's tangent is zeroed.

$$\frac{dy_1}{d\xi} = tg\alpha(\xi) = 0 \tag{43}$$

The inclination angle reach the maximum value 2nd derivative of function y_1 is equal to 0.

$$\frac{d}{d\xi}\left(\frac{dy_1}{d\xi}\right) = \frac{d}{d\xi}\left(\frac{d}{d\xi}\left(\frac{1}{2}\xi^2 - 16\xi^3\right)y_{max}\right) = 96y_{max}\left(\frac{1}{4} - \xi\right) = 0 \tag{44}$$

The maximum takes place at point $\xi_1^* = 0.25$, because $tg\alpha(\xi_1^*)'' < 0$.
For the function y_2:

$$\frac{d}{d\xi}\left(\frac{dy_2}{d\xi}\right) = \left(-4 + 24\xi - 26\xi^2 + 16\xi^3\right)'' \cdot y_{max} = (-72 + 96\xi) \cdot y_{max} = 0 \tag{45}$$

Therefore $\xi_2^* = 0.75$.

For interval $\xi \in \langle 0, \frac{1}{2} \rangle$ and $y_{max} = r_{max}$ based on Eq. 40

$$y(\xi) = 16 \cdot \left(\frac{3}{4} - \xi \right) \cdot \xi^2 \cdot r_{max} \tag{46}$$

while for $\xi \in \langle \frac{1}{2}, 1 \rangle$

$$y(\xi) = \left(-4 + 24\xi - 36\xi^2 + 16\xi^3 \right) \cdot r_{max} \tag{47}$$

therefore

- interval $\xi \in \langle 0, 1/2 \rangle$

$$\frac{dy}{dx} = \frac{dy}{d\xi} \cdot \frac{d\xi}{dx} = \frac{dy}{d\xi} \cdot \frac{d}{dx} \left(\frac{x - x_0}{1} \right) = \zeta \cdot \frac{1}{1} \cdot tg\alpha(\xi), \zeta = \frac{r_{max}}{r_0} \tag{48}$$

- interval $\xi \in \langle 1/2, 1 \rangle$

$$\frac{dy}{dx} = \frac{1}{1} \cdot \frac{dy}{d\xi} = \zeta \cdot \frac{r_0}{1} \cdot tg\alpha(\xi), \zeta = \frac{r_{max}}{r_0} \tag{49}$$

The patch width

- interval $\xi \in \langle 0, \frac{1}{2} \rangle$

$$s(\xi) = 2\pi y(\xi) - 2\pi r_0 = 2\pi(y(\xi) - r_0) = 2\pi \left[16\xi^2 \left(\frac{3}{4} - \xi \right) r_{max} - r_0 \right]$$

$$= 2\pi r_0 \cdot \left[16\xi^2 \left(\frac{3}{4} - \xi \right) \frac{r_{max}}{r_0} - 1 \right] = 2\pi r_0 \left[16\xi^2 \left(\frac{3}{4} - \xi \right) \cdot \zeta - 1 \right], \zeta = \frac{r_{max}}{r_0} \tag{50}$$

- interval $\xi \in \langle \frac{1}{2}, 1 \rangle$

$$s(\xi) = 2\pi y(\xi) - 2\pi r_0 = 2\pi \cdot 4 \left(-1 + 6\xi - 9\xi^2 + 4\xi^3 \right) \cdot r_{max} - 2\pi r_0$$

$$= 2\pi r_0 \left[4 \left(-1 + 6\xi - 9\xi^2 + 4\xi^3 \right) \frac{r_{max}}{r_0} - 1 \right]$$

$$= 2\pi r_0 \left[4 \left(-1 + 6\xi - 9\xi^2 + 4\xi^3 \right) \zeta - 1 \right] \tag{51}$$

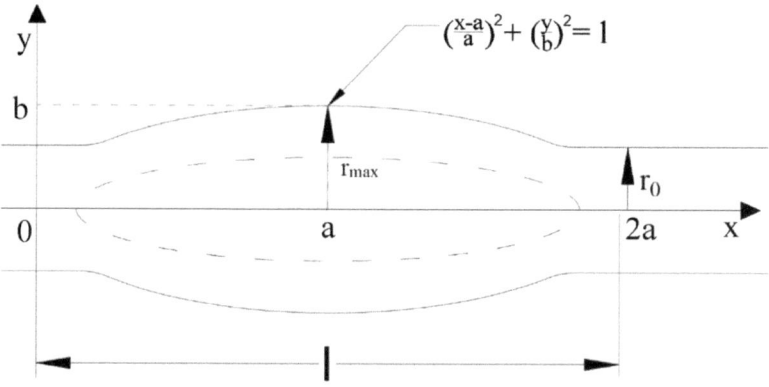

Fig. 10 Elliptical patch

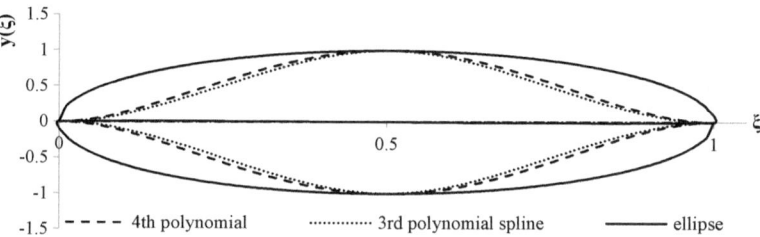

Fig. 11 Comparison of patch geometries

2.1.3 Elliptical Patch

The elliptical equation takes the following form Figs. 10, 11:

$$\left(\frac{x-a}{a}\right)^2 + \left(\frac{y}{b}\right)^2 = 1 \tag{52}$$

Upon introduction of a dimensionless coordinate $\xi = \frac{x}{l} \rightarrow x = \xi \cdot l$ Eq. (1) takes the form

$$\left(\frac{\xi \cdot l - a}{a}\right)^2 + \left(\frac{y}{b}\right)^2 = 1 \tag{53}$$

therefore

$$y = \pm b \sqrt{1 - \left(\frac{l}{a}\xi - 1\right)^2} \tag{54}$$

The perimeter of artery (with radius r_0) with inserted elliptical path is ($l = 2a$)

$$p(\xi) = 2\pi r_0 + 2y = 2\pi r_0 + 2b\sqrt{1 - (2\xi - 1)^2} \tag{55}$$

therefore, the variable artery radius

$$r(\xi) = \frac{p(\xi)}{2\pi} = r_0 + \frac{b}{\pi}\sqrt{1 - (2\xi - 1)^2} \tag{56}$$

and for $\xi = 1/2$

$$r_{max} = r\left(\frac{1}{2}\right) = r_0 + \frac{b}{\pi} \tag{55}$$

For the model case $r_0 = 5$ mm, $b = 4$ m, $l = 40$ mm

$$r_{max} = 5 + \frac{4}{\pi} \approx 6,27 \approx 6,3\,\text{mm} \tag{56}$$

whereas $b = r_{max} - r_0$, therefore

$$y(\xi) = r_0 + (r_{max} - r_0)\sqrt{1 - (2\xi - 1)^2}, \; \xi = \frac{x}{l} \rightarrow d\xi = \frac{1}{l}dx \tag{57}$$

$$\frac{dy}{dx} = \frac{dy}{d\xi} \cdot \frac{d\xi}{dx} = \frac{1}{l}\frac{dy}{d\xi} = \frac{r_{max} - r_0}{l} \cdot \frac{d}{d\xi}\sqrt{1(2\xi - 1)^2}$$

$$= \frac{r_0}{l}\left(\frac{r_{max}}{r_0} - 1\right) \cdot \frac{-2(2\xi - 1)}{\sqrt{1 - (2\xi - 1)^2}} = tg\alpha(\xi), \; \xi \in \langle 0, 1\rangle \tag{58}$$

It is conspicuous that for $\xi = 1$ $\alpha = 90°$. The angle of the tangent to curve $y(\xi)$ decreases from $90°$ to $0°$ in the interval $\xi \in \langle 0, \frac{1}{2}\rangle$ and further grows to $90°$ for $\xi = 1$.
Patch width
The perimeter width of the patch is

$$s(\xi) = 2\pi y(\xi) - 2\pi r_0 = 2\pi r_0(\zeta - 1)\sqrt{1 - (2\zeta - 1)^2}, \zeta = \frac{r_{max}}{r_0} \tag{59}$$

or in the dimensionless form

$$\frac{s(\xi)}{2\pi r_0} = (\zeta - 1) \cdot \sqrt{1 - (2\xi - 1)^2}, \xi \in \langle 0, 1\rangle \tag{60}$$

2.2 Summary

Figure 6 presents a comparison of the patch width curve determined by the function:

- polynomial:

$$y(\xi) = 16\xi^2(1 - \xi)^2, \ \xi \in \langle 0, 1 \rangle \tag{61}$$

- spline:

$$y(\xi) = 12\xi^2 - 16\xi^3, \ \xi \in \left\langle 0, \frac{1}{2} \right\rangle \tag{62}$$

$$y(\xi) = -4 + 24\xi - 36\xi^2 + 16\xi^3, \ \xi \in \left\langle \frac{1}{2}, 1 \right\rangle \tag{63}$$

- elliptical:

$$y(\xi) = \sqrt{1 - (2\xi - 1)^2}, \ \xi \in \langle 0, 1 \rangle \tag{64}$$

2.3 Numerical Analysis

To obtain a full description of fluid flow kinematics, a system of equations is solved describing the principles of conservation of momentum and mass, which form the basis to calculate velocity, pressure, density and temperature. No analytical solution of the general Navier–Stokes equation has been found, therefore the determination of an accurate solution is impossible from the analytical point of view. However, satisfactory solutions are obtainable approximating the solution of the equation by numerical methods. In the case of the research carried out herein, this was the finite volume method [7, 8, 9].

To describe a variation of any physical value in space and time, transport equations are used, also referred to as the conservation equations. To describe blood flow in the arteries, two equations are used:

- momentum transport equation:

$$\rho \left(\frac{\partial u_i}{\partial t} + u_j \frac{\partial u_i}{\partial x_i} \right) = -\frac{\partial p}{\partial x_i} + \mu \frac{\partial^2 u_i}{\partial x_j \partial x_j} \tag{65}$$

- mass transport equation:

$$\frac{\partial u_i}{\partial x_i} = 0 \tag{66}$$

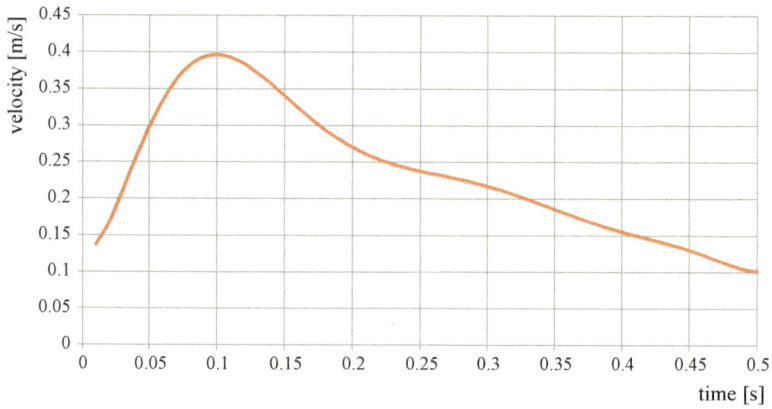

Fig. 12 Inlet velocity profile (one pulse duration $= 0.5$ s)

After the numerical solution of the above equations, the components of velocity and pressure are obtained, which enables the acquisition of full knowledge of the flow nature. The system of equations described above, due to the assumption of averaged velocity value (Reynolds distribution), must be supplemented with additional equations considering the velocity fluctuations. The additional system of equations is called the turbulence model. The k-ω SST model was selected for the research, due to its versatility and efficient detection of separations of the wall layer [9, 10]. An additional parameter determined by the said equations is the kinetic energy of turbulence defining the energy of whirlpools occurring in the flow. The larger and stronger the whirlpools, the higher the value of the kinetic energy of turbulence. Thanks to this property, this parameter is the foundation for the arterial flow assessment in terms of turbulence.

A nonstationary, pulsating velocity profile was applied at the artery inlet. Figure 12 presents the course of one cycle, which takes 0.5 s. The selected velocity profile was prepared based on ultrasound test results and data found in relevant literature [11, 12]. The maximum velocity is 0.4 m/s. The curve presenting velocity was shown in Fig. 12. The curve is described by the polynomial function. At the outlets, the zero-pressure boundary condition was applied [12, 13]. This is determined by the fact that the primary purpose of the study is to define the impact of the geometry on the flow. The introduction of the pulsating outlet pressure profile could have caused some additional disorders, physically unrelated to the purpose of this study. In order to examine the impact of the velocity on the flow, a series of simulations were made for the pulsating velocity profile of the maximum value of 1 m/s. Its curve is identical and the only difference is the amplitude.

Blood is a liquid of a viscosity highly dependent on the velocity of the flow. This dependency is not linear, therefore blood is classified in the non-Newtonian group of fluids. At low velocities the viscosity value is very high. It enhances the proneness of the wall layer to separate as well as the occurrence of whirlpools compared to the

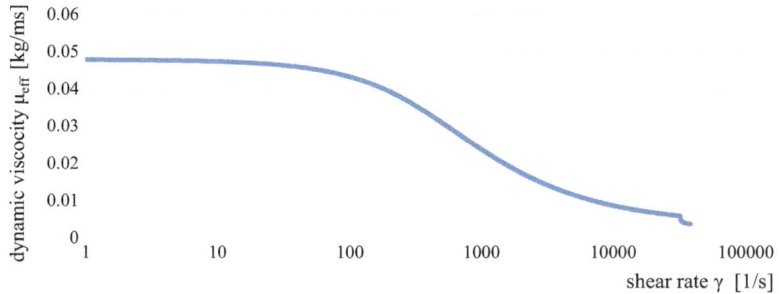

Fig. 13 Carreau function

Newtonian fluids such as water [14, 15]. In the course of the research, the Carreau model [16, 17, 18, 12]. The Carreau formula is presented by Eq. 67. In Fig. 13, change of dynamic viscosity as a function of shear rate is presented. The coefficients in Eq. 67 were selected based on the experimental tests [17].

$$\mu_{\text{eff}}(\dot{\gamma}) = \mu_\infty + (\mu_0 - \mu_\infty)\left(1 + (\lambda\dot{\gamma})^2\right)^{\frac{n-1}{2}} \tag{67}$$

of which:

$\dot{\gamma}\left[s^{-1}\right]$—deformation of velocity given by the formula:

$$\dot{\gamma} = \frac{\partial u_i}{\partial x_i}\tau^{-1} \tag{68}$$

$\lambda = 3.13$ s—fluid relaxation time

$n = 0.3568$—exponent

$\mu_\infty = 0.00345$ kg/m·s—viscosity value at indefinitely high coagulation velocity (in the flow core),

$\mu_0 = 0.056$ kg/m·s—viscosity value at zero coagulation velocity (near the walls).

3 Results

Based on the equations obtained in the analytical parts, several geometries have been produced. The obtained models can be divided into three groups, differing by artery diameters. In each group, the artery expansions from 10% of the original dimension to 50% were examined. The examinations were made for unilateral expansion of the patch. Figure 14 presents the results of the simulation for the same expansion level, with a symmetrical expansion in the first case and unilateral in the second. It is noticeable that in the nonsymmetrical case the created whirlpool is larger, which makes the plaque deposition more probable. With the symmetrical expansion, the whirlpools are smaller, because they occur in two locations. The conclusion of such

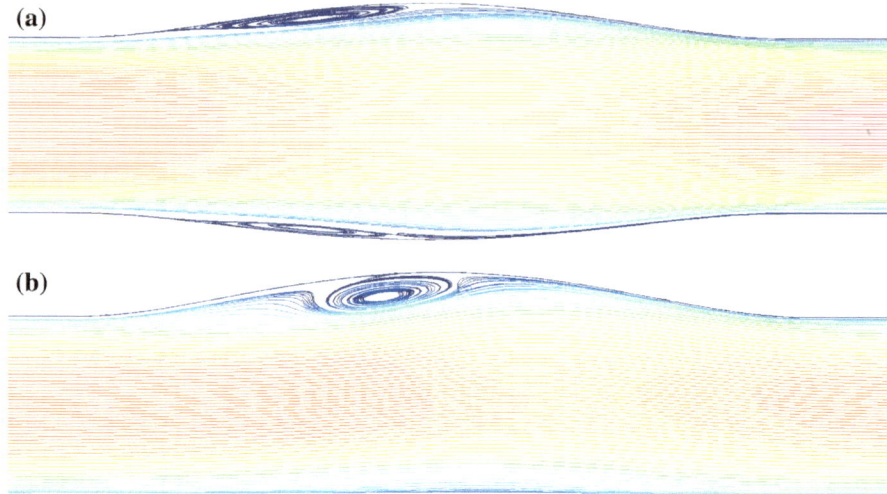

Fig. 14 Dilation of the artery: **a** symmetric **b** asymmetric

a comparison is that cases of unilateral expansion are worse in terms of the flow. They also occur more frequently in practice, particularly when the patch is inserted near the carotid artery bifurcation area, which is the major object of the study.

Figure 15 presents the curve of the current line for the patch described by grade 4 polynomial and the patch described by the function composed of grade 3 polynomials (spline). It can be observed that the curve of the current line is practically the same. Therefore, only one geometry was examined (function composed of third-grade polynomials—due to its higher accuracy).

The geometry defined by the elliptical function is the most approximate to the patches inserted presently. Figure 16a presents the current lines for the elliptical case. The graphical presentation Fig. 16b shows the results for the polynomial patch with the same maximum expansion radius. The conclusion of the drawing can be that the flow through the elliptical geometry causes whirlpools already at a 20% expansion, while no whirlpools occur in the elliptical patch. This is the main reason for the application of polynomial patches.

Based on the above preliminary studies, the authors chose to carry out a detailed examination of unilateral expansions occurring upon the insertion of the patch defined by a polynomial function. For each group of geometries, the value of the coefficient of stenosis was determined defined by the ratio of the maximum radius of an expanded artery to the radius of a normal artery, thanks to which the parameter examined can be described in a dimensionless way. Figure 17 presents the streamlines in the diastolic phase where the negative velocity gradient impacting the highest probability of boundary layer separation.

A similar analysis was made for the other groups. All the results have been included in Table 1. It is noticeable that for lower velocities, the use of a narrower

(a) 4th polynomial

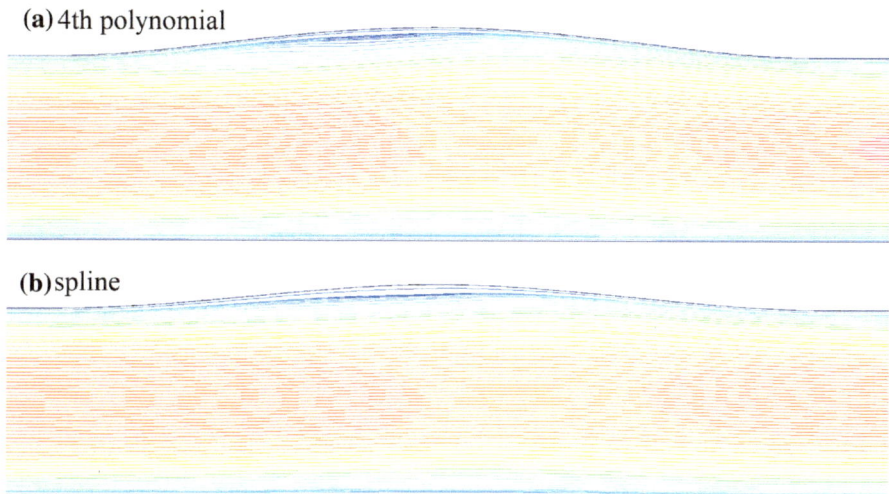

(b) spline

Fig. 15 Comparison of geometries: **a** fourth-degree polynomial **b** third-degree polynomial spline

(a)

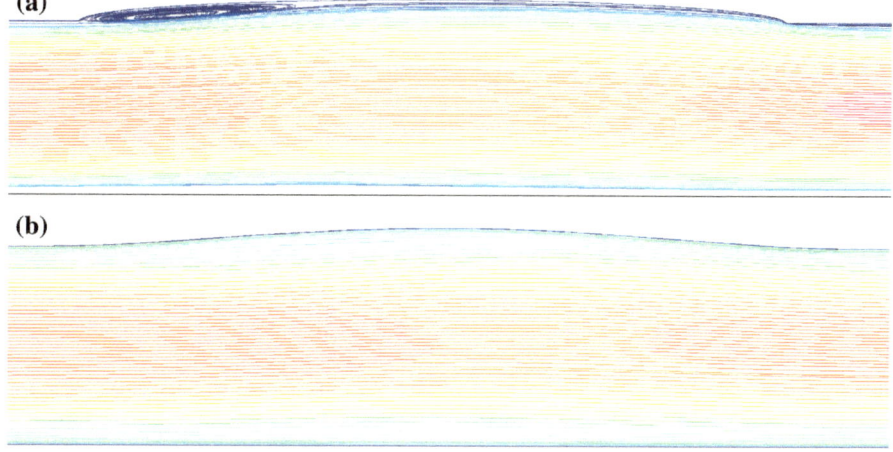

(b)

Fig. 16 Comparison of the artery geometry with the patch sewn in in the shape of **a** ellipse **b** polynomial

patch is required. This is due to the fact that slower blood flow enhances the growth of the viscosity forces. The higher viscosity forces in the fluid, the higher the tensions coagulating the fluid layers, which causes higher inclination of the fluid to separations and whirlpools. The final results assumed in the studies comprised the patch widths of lower velocity values, because they constitute a higher safety limit. The blood flow velocity, most of all, depends on the patient's pulse, artery diameter, and the difference in the value of the systolic and diastolic blood pressure. This value is variable, therefore the worst-case scenario method was applied (Fig. 18).

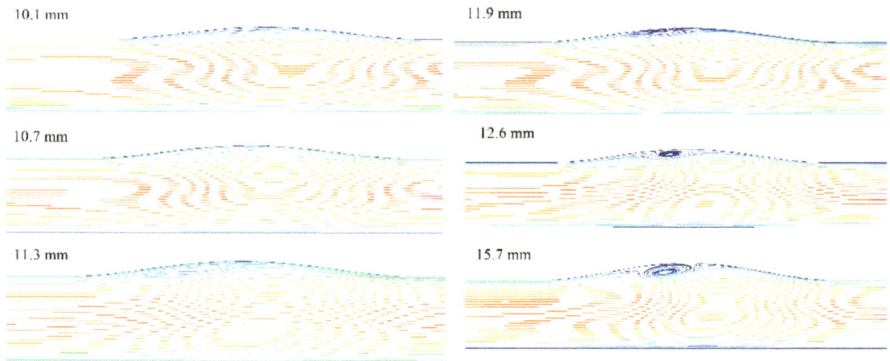

Fig. 17 The course of streamlines depending on the width of the patch [mm]

Table 1 Numerical results

Artery diameter (mm)	Velocity (m/s)	4th polynomial	3rd polynomial spine	Ellipse
8	0.4	10.6	10.6	12
	1	10.1	9.6	
10	0.4	12.6	11.3	
	1	11.3	10.1	
12	0.4	14.3	13.6	
	1	12.8	12.1	

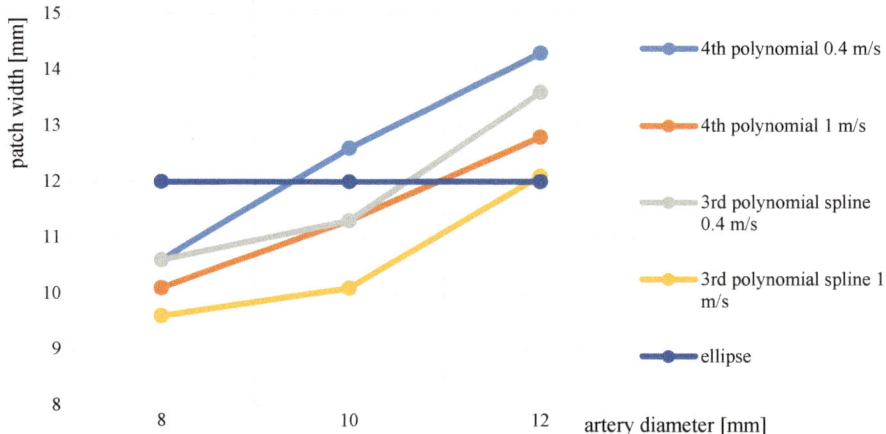

Fig. 18 Limit width of the patch depending on the diameter of the artery (velocity 0.4 m/s; 1 m/s)

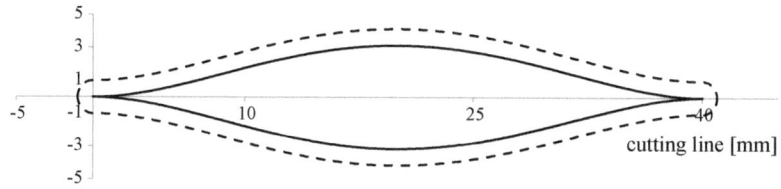

Fig. 19 Patch outline with overlap on the stitching (external profile) and without overlap on the stitching (internal profile), for the cutting length $l = 40$ mm, the width of the overlap on the suture $s = 1$ mm

The obtained values represent the starting points with a view to continuing and expanding the research. The final result will be the development of a complete patch geometry (Fig. 19), the insertion of which in the patient's artery would minimize the risk of plaque redeposition. The finished patch will be increased by 1 mm regarding the suturing technique. It will be customized to each individual patient. The studies carried out enabled the authors to determine the first input parameter: the patient's artery diameter. Finally, there will be more input parameters and they will consider the following: blood composition, carotid artery bifurcation angle, diameters of adjacent arteries (usually the main CCA and external ECA).

4 Conclusions

The performed analysis has shown that the most favorable geometry in terms of the flow is the artery expanded according to a polynomial function. Thanks to the appropriate formation of the walls, the lumen expands so gradually that the boundary layer increases slowly. While increasing the patch width, the risk of larger separations and reverse flow occurs. The larger the diameter of the patient's artery, in which the patch will be inserted, the larger the width that can be used (not exceeding 15 mm). Theoretically, the smaller the patch the better, since the artery geometry is not significantly changed. If, however, the patch sutured in is too small as a result of the patch shrinking, it may lead to a reduction of the arterial lumen. It is a more negative phenomenon than the expansion, because it restricts the blood flow. If plaque redeposition occurs in such an artery, it could be closed much sooner than with an expanded artery. Therefore, the information related to the width of the patch that can be applied is very important to the surgeons. The research allowed the authors to obtain widths that are justified analytically and mechanically.

The patches inserted thus far were similar to an ellipse or were of an irregular shape. They met their purpose, however, significantly affected the blood flow, which in some cases led to the necessity of subsequent surgery. The patches suggested in this paper have two basic advantages. The first is that the geometry of the patches has been described analytically, which enables their easy recovery with the use of simple aids. The other, more important aspect is the standardization of the geometry of the inserted patches. The patches can be divided into several types adapted to the majority of patients and to the most commonly occurring diameters. Thanks to such a solution, the patches can be mass-produced in several variants. During the medical procedure, the surgeon would select a finished, ready-cut patch, depending on the patient's geometry. Hence, there would be no need to cut the patches while operating, which would reduce the time of the procedure and, most importantly, minimize the risk of restenosis.

References

1. May AG, Van de Berg L, Deweese JA, Rob CG (1963) Critical arterial stenosis. Surgery 54(1):250–259
2. Smoore W, Malone JM (1979) Effect of flow rate and vessel calibre on critical arterial stenosis1. J Surg Res 26 (1979)
3. Berguer R, Hwang NHC (1974) Critical Arterial Stenosis: A Theoretical and Experimental Solution. Ann Surg 180(1):39
4. Liapis CD, Bell SPRF, Mikhailidis D, Sivenius J, Nicolaides A, Fernandes e Fernandes J, Biasi G, Norgren L (2009) ESVS guidelines. invasive treatment for carotid stenosis: indications, techniques. Eur J Vasc Endovasc Surg
5. Muto A, Nishibe T, Dardik H, Dardik A (2009) Patches for carotid artery endarterectomy: current materials and prospects. J Vasc Surg 50(1):206–213
6. Rerkasem K, Rothwell PM (2009) Patch angioplasty versus primary closure for carotid endarterectomy. Cochrane Database Syst Rev
7. Chung TJ (2009) Computational fluid dynamics
8. Schäfer M (2006) Computational engineering: introduction to numerical methods
9. ANSYS: Ansys Fluent 14.0 tutorial guide (2009) Ansys INC
10. Botar CC, Vasile T, Sfrangeu S, Clichici S, Agachi PS, Badea R, Mircea P, Cristea MV (2010) Validation of CFD simulation results in case of portal vein blood flow. Comput Aided Chem Eng
11. Moyle KR, Antiga L, Steinman DA (2006) Inlet conditions for image-based CFD models of the carotid bifurcation: is it reasonable to assume fully developed flow? J Biomech Eng 128:371
12. Moon JY, Suh DC, Lee YS, Kim YW, Lee JS (2014) Considerations of blood properties, outlet boundary conditions and energy loss approaches in computational fluid dynamics modeling. Neurointervention
13. Steinman DA, Robarts JP, Milner JS, Moore JA, Rutt BK (1997) Hemodynamics of human carotid artery bifurcations: computational studies with models reconstructed from magnetic resonance imaging of normal subjects
14. Chien S, Usami S, Dellenback RJ, Gregersen MI (1967) Blood viscosity: influence of erythro-cyte deformation. Science (80) 157:827–829
15. Chen J, Lu X-Y, Wang W (2006) Non-Newtonian effects of blood flow on hemodynamics in distal vascular graft anastomoses. J. Biomech. 39
16. Khan MF, Quadri ZA, Bhat SP (2013) Study of Newtonian and non-Newtonian effect of blood flow in portal vein in normal and hypertension conditions using CFD technique. Int J Eng Res Technol 6:974–3154
17. Boyd J, Buick JM, Green S (2007) Analysis of the Casson and Carreau-Yasuda non-Newtonian blood models in steady and oscillatory flows using the lattice Boltzmann method. Phys Fluids
18. Siebert MW, Fodor PS (2009) Newtonian and Non-Newtonian blood flow over a backward-facing step—a case study. In: Proceedings of the COMSOL Conference 2009, Boston

Numerical Study of Carotid Bifurcation Angle Effect on Blood Flow Disorders

N. Lewandowska, M. Micker, M. Ciałkowski, M. Warot, A. Frąckowiak and P. Chęciński

Abstract The paper presents the study results of the impact of the common carotid artery bifurcation angle on the flow disorders. The studies were carried out using numerical methods. Based on actual images, geometry was made of the diffuser channel with bifurcation and predetermined angle. The flow simulation results showed that for bifurcation angles exceeding 60° the vortices near the bulb start to occur—at that time almost a double increase of the parameter values takes place, related to flow disorders. The vortex becomes increasingly larger and grows proportionally to the value of the bifurcation angle. Thanks to the studies carried out, three areas have been shown, in which plaques may deposit, due to disadvantageous geometry. Based on the simulation results, arteries have been divided into three groups of risk. It has been proven that bifurcations exceeding 50° significantly disturb the flow and the points of whirlpool occurrence represent frequent points of plaque depositions.

1 Introduction

This paper presents the study results aimed at the determination of anatomy of carotid arteries with geometry enhancing plaque deposition. The application of numerical methods in medicine becomes increasingly popular. There are vascular surgery issues that, in addition to biological (physiological), have a mechanical (hemodynamic) cause. Thanks to the combination of the two fields of science, the phenomena occurring in the artery can be explained more completely. Through analysis of the flow field in the common carotid artery bulb, the phenomenon of boundary layer separation and formation of vortexes can be detected. If the vortex occurring in the flow is

N. Lewandowska (✉) · M. Ciałkowski · A. Frąckowiak · P. Chęciński
Faculty of Machines and Transport, Chair of Thermal Engineering, Poznan University of Technology, Piotrowo 3, 60-695 Poznan, Poland
e-mail: natalia.lewandowska.pp@gmail.com

M. Micker · M. Warot · P. Chęciński
Department of General and Vascular Surgery and Angiology, Poznan University of Medical Sciences (PUMS), 34 Dojazd St, 60-631 Poznan, Poland

© Springer Nature Switzerland AG 2019
J. M. R. S. Tavares and P. R. Fernandes (eds.), *New Developments on Computational Methods and Imaging in Biomechanics and Biomedical Engineering*, Lecture Notes in Computational Vision and Biomechanics 33, https://doi.org/10.1007/978-3-030-23073-9_2

Fig. 1 Deposition of solid particles following the vortex formation

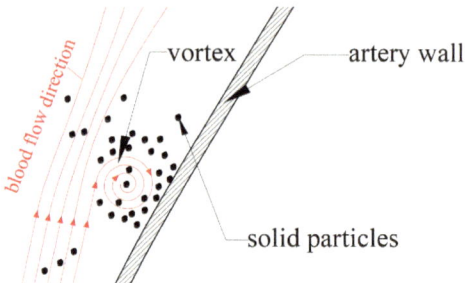

sufficiently large and strong, it causes a "suction" of solids present in the blood such as components of atherosclerotic plaque. As a result, the said particles deposit on the arterial walls. They will cause a progressive stenosis, which enhances the process of further formation of vortexes (Fig. 1). This is one of the theories of atherosclerotic plaque formation in the point of arterial bifurcation.

The literature [1–6] concerning numerical studies of carotid arteries presents studies mainly focused on carotid artery thicknesses and diameters of the cross-section of its branches. The analysis was mainly carried out through presentation of the shearing stress field and the velocity values in the canal. It was shown that arterial narrowing leads to increased blood flow velocity and, as a consequence, growth of the shearing stresses, which enhances flow turbulization and affects the formation of deposits. In addition, weakening of the arterial walls caused by lesions in the patient causes reduced flexibility of the wall and local expansion of internal carotid artery, which also causes deposition of atherosclerotic plaques and may also cause their formation. The deposits most often occur on the side of the internal carotid artery, which is particularly dangerous, because this artery transports blood to the brain and its closing very often causes irreversible damage to the cerebral tissue or even death.

The formation of deposits on the side of the internal carotid artery is, most of all, enhanced by its geometry—the increasing cross-section of the flow causes a reduction of the blood flow and, as a consequence, its increased viscosity, which has been explained in detail further in the paper. High viscosity enhances the formation of deposits, because the separation of the boundary layer occurs near the walls and vortexes occur more easily. Except for the aspects directly concerning the geometry of the specific arteries, the parameters correlating the external and internal carotid artery are also significant.

Bulb is the key region, in which the blood flow is most exposed to disorders— this is the area where the artery expands and is bifurcated into two smaller arteries (Fig. 2).

In contrast to the previous analyses in this field, the authors decided to focus on one though very important parameter, differentiating the carotid arteries. This is the bifurcation angle of the common carotid artery. This paper will include studies on the impact of the change of the said angle on the blood flow parameters. It is rarely considered in blood flow analyses, though it expressly enables the determination of some groups of carotid arteries with geometry enhancing the formation of deposits.

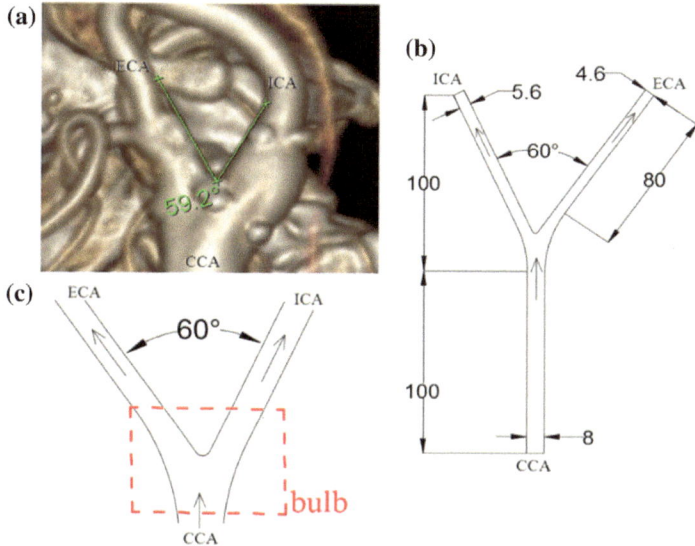

Fig. 2 Carotid artery: **a** real **b** geometric model **c** bulb

From the flow point of view, the area distinctly changes such parameters as flow velocity, pressure and, most importantly, stability. The purpose of the studies was to check, with the use of numerical methods, whether vortexes would occur with correspondingly large bifurcation angles, which, as a consequence might lead to the formation of deposits. The purpose of the study was to determine geometrically correct arteries. The practical result of the studies is the possibility to apply proactive measures aimed at the delay or reduction of the risk of deposits formation in patients with arteries classified within the risk group.

2 Methods

The studies were carried out using the ANSYS software, through numerical modeling of the flow. The CFD (*Computational Fluid Dynamics*) tools enable a very good reproduction of the flow, if the appropriate number of parameters is included in the model. Figure 2 presents a general outline of geometry. The CCA (*Common Carotid Artery*) was treated as a nonsymmetric diffuser canal. Based on the analysis of MRI images of arteries and papers treating on carotid artery studies [1–6], it was observed that the bifurcation of the ICA (*Internal Carotid Artery*) is by ca. 5°–10° larger than the bifurcation of the ECA (*External Carotid Artery*). The ICA and ECA were also approximated as straight-line canals. The main area of study—the bulb, was distinguished in Fig. 2c. The increased length of the canals results from the fact that laminar stabilized inlet and outlet flows were targeted during the simulation.

Analytical model of the artery is described by a system of three or four equations, depending on the number of dimensions. Those equations are called momentum transport equation and continuity equations. The solution of this system gives values of velocity and pressure in every node of the created mesh.

As a consequence of using Reynolds decomposition in determining the values, in mathematical models appears Reynolds Stress Tensor (RST). It includes the fluctuation of physical quantities over time. The solution of RST demands direct RST solving, which adds an extra seven equations to the model [7, 8]. It significantly increases the time of calculation. Instead of implementing all RST to the analytical model, two-equation turbulence model was chosen. It gives high accuracy in detecting flow disturbances. In the next paragraph, the selected model will be described more precisely. Summarizing, analytical model is described by 5 (or 6 for 3D geometry) equations:

- momentum transport equation:

$$\rho\left(\frac{\partial u_i}{\partial t} + u_j \frac{\partial u_i}{\partial x_i}\right) = -\frac{\partial p}{\partial x_i} + \mu \frac{\partial^2 u_i}{\partial x_j \partial x_j} \tag{1}$$

- mass transport equation (continuity equation):

$$\frac{\partial u_i}{\partial x_i} = 0 \tag{2}$$

- equations connected with Reynolds stress tensor (depending on turbulence model), in considered case: 2 equations.

The most important aim of the studies was the watch of the boundary layer separation and the formation of the vortex. The accuracy of detection of the said phenomena is determined by the turbulence models. For each case, depending on the flow conditions and parameter values, an individual model should be selected. After a detailed analysis, the k-ω SST model was selected, which very effectively reproduces the points of the boundary layer separation [9].

The boundary conditions define the velocity or the pressure values in the artery flow inlets and outlets. The velocity value was selected based on actual values occurring in the common carotid arteries [5, 10, 11] equaling to 1 m/s. At the arterial flow outlets, the zero-pressure boundary condition was applied to settle the zero overpressure value at the outlets. The condition is particularly often applied in the cases where the most important study parameter is the nature of the flow [5, 11].

In reality, the arterial blood flow is of a pulsating nature. In the studies, a stationary flow with constant-velocity profile at the inlet (constant in time) was applied. The simplification is reasonable because the authors' main focus was the impact of the geometry on the nature of the flow.

The blood flowing through the carotid arteries is an untypical fluid, referred to as the non-Newtonian fluid. It differs from Newtonian fluids by the fact that its viscosity is not a constant value. Viscosity is a physical value included in the description of a relation between the shearing stresses occurring in fluid τ and the fluid veloc-ity gradient u along normal direction to the surface of occurrence of the shearing stresses x_i (referred to as the non-dilatational shear rate) [7]. High viscosity fluids are characterized with the fact that with low velocities even the boundary layer might be separated and strong flow disorders might occur. Blood is a pseudo-plastic fluid and its viscosity reaches high values at low velocity, however, the faster it flows in the artery the lower the viscosity. The Carreau model [12–15] was used to create the blood model (Eq. 3). In contrast to other, less complex models, it implements the value referred to as the fluid relaxation time, which causes a delay of viscosity drop as a result of the strain velocity growth and is milder, which reflects the nature of the pseudo-plastic fluid and the viscosity value in the area of low shearing velocities. It leads to higher effectiveness of detection of the boundary layer separations.

The values of the coefficients in Eq. 3 were selected based on the paper, in which they were determined following a series of experiments [12]:

$$\mu_{\text{eff}}(\dot{\gamma}) = \mu_\infty + (\mu_0 - \mu_\infty)\left(1 + (\lambda\dot{\gamma})^2\right)^{\frac{n-1}{2}} \tag{3}$$

of which:

$\dot{\gamma} = \frac{\frac{\partial u_i}{\partial x_i}}{\tau}\left[\frac{1}{s}\right]$—shear rate
$\lambda = 3.13$ s—fluid relaxation time
$n = 0.3568$—exponent
$\mu_\infty = 0.00345$ kg/m·s—viscosity value at infinitely high shearing velocity (in the flow core),
$\mu_0 = 0.056$ kg/m·s—viscosity value at zero shearing velocity (near the walls).

Upon the analysis of the actual images of the arteries, the range of angles between the external and internal carotid arteries occurring, in reality, was determined. The value of the smallest bifurcation angle is 15°, while the largest bifurcation detected was 90°. To analyze the flow at various angles, 9 models of the following angle values have been made: 15°, 30°, 40°, 45°, 50°, 55°, 60°, 75°, and 90°. It was observed that from 40° to 60° the angle was changed every 5°, because, in this area, the first larger separations were expected.

For the purpose of comparing 2D results with their equivalent of 3D, three 3D models were made. The models have the same diameters and lengths as 2D models. Considered angles of bifurcation in cases of 3D calculations was 25°, 50°, and 75°. The meshes were made in ICEM. For the 2D geometry, the fully structured mesh was made. For 3D models, triangular mesh was generated with extra prism-layers near the walls to improve mesh quality and increase accuracy for boundary layers. The meshes were shown in Fig. 3.

To assess the flowing nature, the authors considered the following factors while developing the results:

- boundary layer thickness in the diffuser part and at the outlets δ [mm]: the value was assessed by determining the areas, in which the fluid velocity was lower than in the flow core. In Fig. 3, a concept of measurement was presented based on the

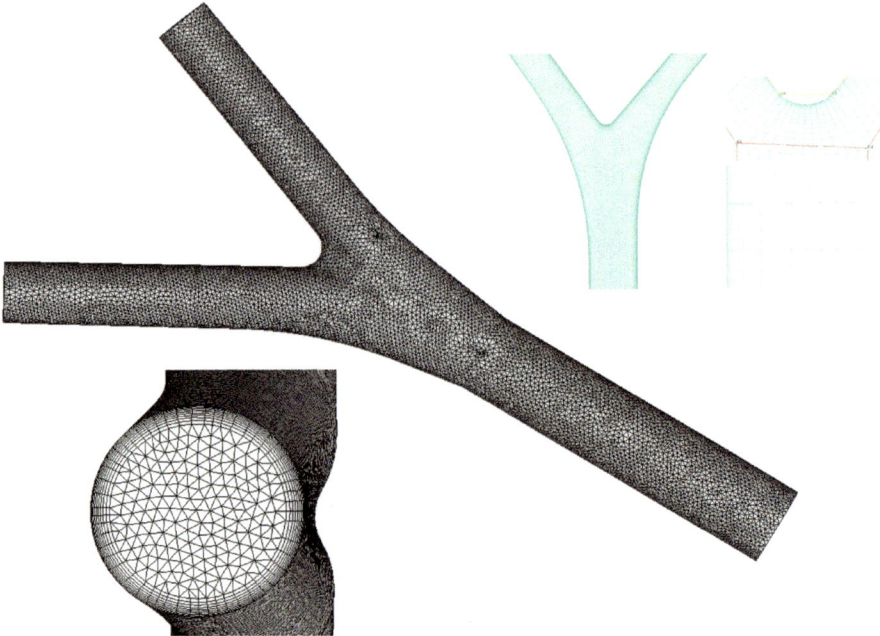

Fig. 3 Computational grid for 3D and 2D geometry

example of one of the outlets. According to the common assumption, the boundary layer occurs until the flow velocity reaches 99% of the core flow value; (Fig. 4)

- turbulent Reynolds number Re_{turb} [-], which defines the turbulent viscosity (defined by Boussinesq function) to laminar viscosity ratio [9]. The probability of occurrence of vortexes and separations grows along with the value of the turbulent Reynolds number in a specific area;

- kinetic energy of turbulence k [J/kg]—average kinetic energy referred to the mass unit and related to the vortexes in the turbulent flow. When the flow becomes turbulent, the share of diffusion in the transport grows. It was assumed that the turbulent flow structure contains areas of highly nonstationary nature, referred to as vortexes. The intensification of vortexes increases the transportation of volumes by diffusion—they can be small or large. The kinetic energy of turbulence is the energy of vortexes in a turbulent flow—the larger they are the more their kinetic energy grows. It is "supplied" from the kinetic energy of flow to larger vortexes, from larger to smaller to dissipate in the end—until the vortexes are so small that the viscosity forces overcome the inertia forces (correlated with kinetic energy). When assessing the field of the kinetic energy of turbulence, one can locate the area of the vortex occurrence and by referring to the value of the said energy to the kinetic energy of fluid, one can estimate what part of the kinetic energy related to laminar flow is transformed into turbulence energy.

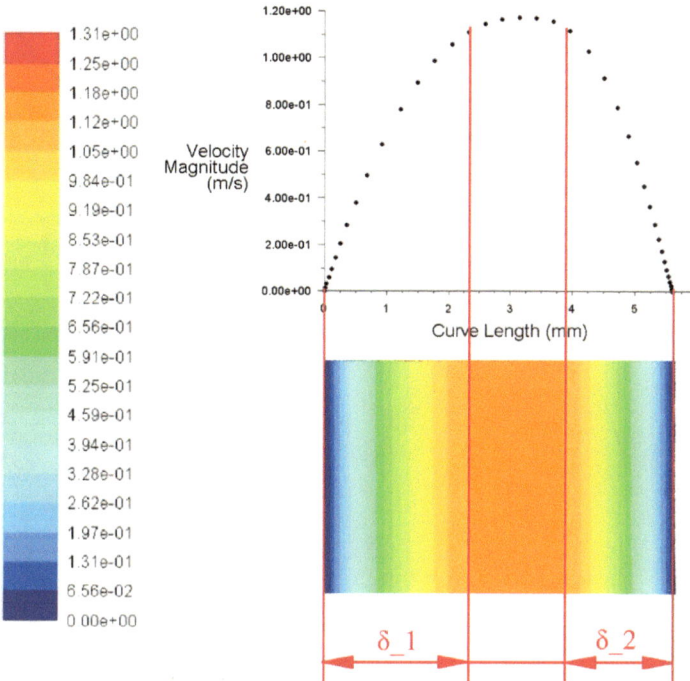

Fig. 4 Measurement of the thickness of the boundary layer

3 Results

Three possible areas occur in the artery, in which, due to the reduction of velocity (i.e., also increase of viscosity) deposits in the form of atherosclerotic plaques may occur. Figure 5 shows the field of turbulent Reynolds number Re_{turb}, which defines the ratio of turbulent viscosity (defined by Boussinesq function) to laminar viscosity [7]. The probability of vortexes and separations grows along with the turbulent Reynolds number in a specific area. The graphical presentation precisely shows that the areas of high Re_{turb} correspond to the locations of deposits. For the smallest angle, the maximum value of Re_{turb} is equal to ca. 3.2. For the angle of 55° maximum value of Re_{turb} significantly increases to 4.5 and the area of high values of Re_{turb} starts to "move" into the walls. For the angle of 75° Re_{turb} has the highest value of 7.45. As we can see in Fig. 5, Re_{turb} increases significantly near the walls—it means that the separation of the boundary layer has begun and flow in this region is severely disturbed.

Allowing for the kinetic energy of turbulence k in the cases presented in the analysis, the areas of the highest value of this energy occur in the bulb, in the vicinity of the walls (Fig. 6). For the smallest angle of 30°, the energy is $0.01537\frac{J}{kg}$, while for the angle of 75°, it is $0.0237\frac{J}{kg}$, i.e., almost double growth of kinetic energy

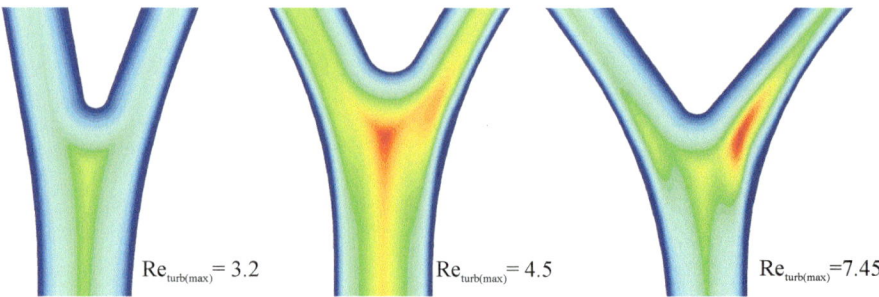

Fig. 5 Distribution of turbulent Reynolds number for 30°, 55°, and 75° bifurcation angle

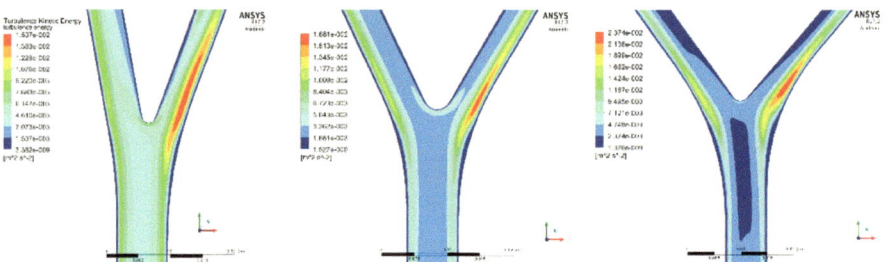

Fig. 6 Turbulence kinetic energy distribution k [J/kg] angle of 30°, 55°, and 75°

of turbulence is detected. In the case of bigger angles, a strong growth around the external artery wall takes place, because the vortex is formed at this point. Distribution of kinetic energy of turbulence is more concentrated near the walls for the highest angle. It is connected with the phenomena of vortex formation—for the high angles, as it will be shown later, vortex are faster, the centrifugal force which is responsible for "suction" of particles is actually bigger, so vortex energy has bigger value and is more focused.

The greatest thickness of the boundary layer occurs in the area, in which the straight-line artery starts to diverge. The increase of the cross-section, which the fluid flows through, causes a drop of pressure, which enhances the separation of the boundary layer. Along with the angle growth, the thickness of the boundary wall grows. Figure 6 presents the field of velocity for the bifurcation angle of 15°, 50°, and 90°. It was observed that the angle growth causes the growth of the separation area, which is the place of the deposit formation, especially on the side of the internal artery (Fig. 7).

From the angle of 60° in the area of the increase of the cross-section, low-velocity vortexes appear, which is shown by the curve of the current line (Fig. 8). It was observed that the vortexes did not occur (or are very small and their velocity is equal to 0) at the angle of 55° and, upon changing of the bifurcation angle by 10°, a rapid growth of the kinetic energy of turbulence (from 0.0105 to 0.016 J/kg), turbulent Reynolds number (from 4.33 to 7.16), and the boundary layer thickness

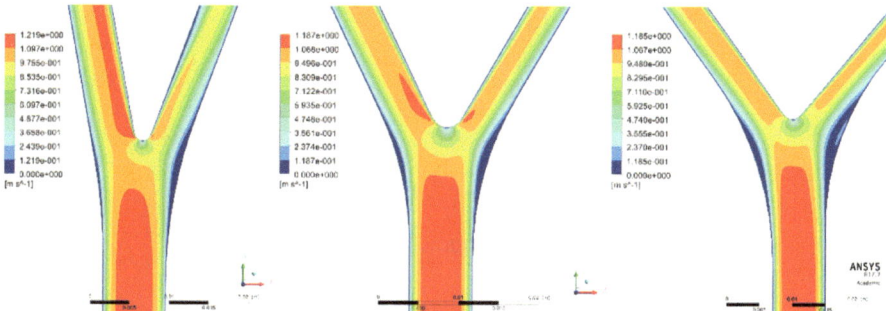

Fig. 7 Velocity distribution for the angles of 30°, 55°, and 75°

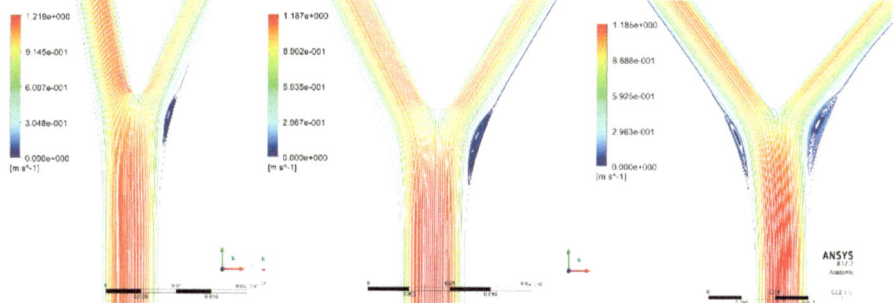

Fig. 8 Pathlines of flow for angle of 30°, 55°, and 75°

(from 2.5 mm do 3.7 mm) were observed. Such flow retreat, even in the low-velocity area, is very dangerous, because it enhances the deposition of atherosclerotic plaques and other solid particles near the walls. The larger the angle, the larger the area of the formed vortex (Fig. 8), and, as a consequence, the flow structure becomes increasingly distorted. In addition, with reduced velocity, the viscosity of blood as a fluid thinned with shearing is very high, which impedes the laminar flow through the arteries. These two factors—the boundary layer and the increased viscosity have the greatest impact on the nature of the flow.

Figure 9 presents the velocity profiles at the outlet from the ICA and ECA and the profile of velocity distribution in the ICA with the angles of 15° and 90°. The nonsymmetrical nature right at the outlet from the bulb results from the earlier flow deceleration at the stagnation point. At the angle of 90°, a separation of the boundary wall occurs along with a clear impact of the stagnation point on the maximum value of velocity in the diffuser part of the artery.

3D Results

To support results obtained for 2D geometries, the calculations for three 3D geometries were made. Comparing the 2D results with outcomes for 3D geometry, it can be observed the same physical regularities. In Fig. 10 distribution of kinetic energy of turbulence was shown for angles of 25°, 50° and 75° (in the axial plane). In this

Fig. 9 Velocity profiles
a angle of 15° **b** angle of 90°

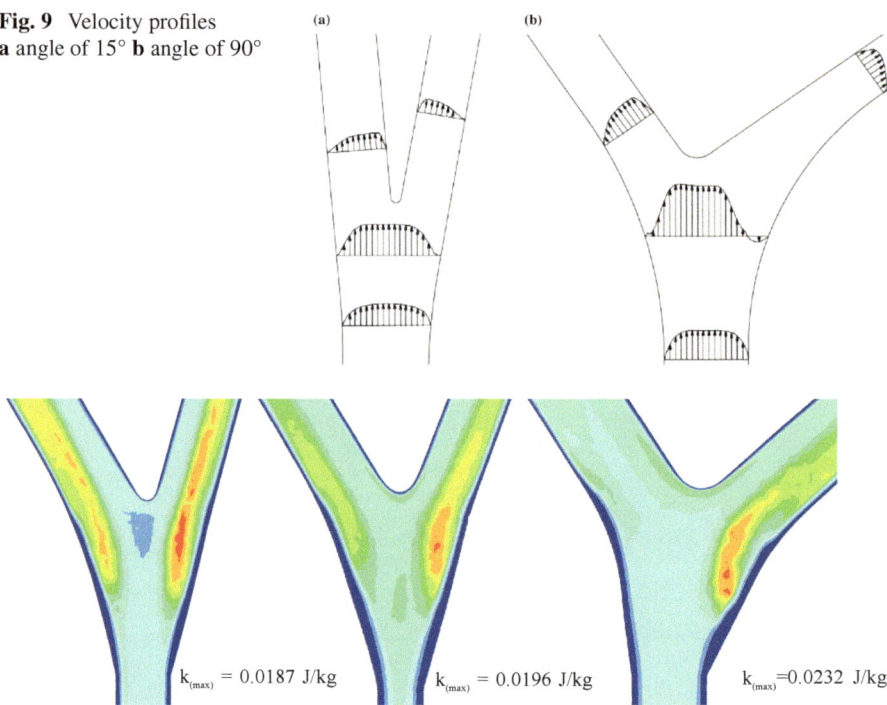

(a)   (b)

k$_{(max)}$ = 0.0187 J/kg   k$_{(max)}$ = 0.0196 J/kg   k$_{(max)}$=0.0232 J/kg

Fig. 10 Turbulence kinetic energy distribution k [J/kg] in the axial plane of 3D artery for angles of 25°, 50°, and 75°

case, it can also be noted the process of increasing concentration of this energy near the walls. This energy is more focused for the widest angles, because of the vortex formation in this part of the artery.

Similarly to 2D geometry, if we consider velocity distribution (Fig. 11) in the artery, for the angle of 75° a development of boundary layer is very prominent—in this part of the bulb, the vortexes start to occur. The interesting part is that the boundary layer in the bulb with a widening angle of 25° and 50° are comparable: the biggest difference is the velocity profile ECA.

In considered cases, very important issue in the carotid arteries is the tendency to vortex formation near the wall. In Fig. 12 this situation was shown (bifurcation angle of 75°). 2D results show with good accuracy the location of vortex formation in the bulb. But if we compare this with 3D results for geometry with the same angle, it can be seen that the character of the vortex is very different from this, what it was obtained from 2D results. The vortex is small and multiplies toward the external part of the wall of the branched artery. Although the vortex occurring in results for 2D geometry is not the way it looks like in practice, these results show also correctly place of vortex formation what is sufficient for the purpose of this research.

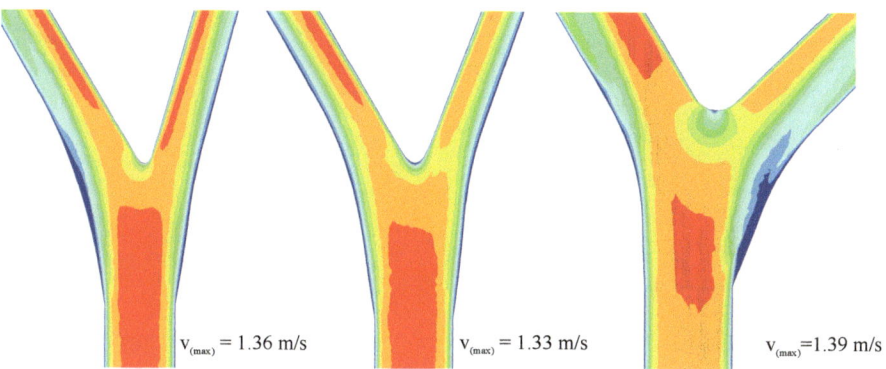

$v_{(max)} = 1.36$ m/s $v_{(max)} = 1.33$ m/s $v_{(max)} = 1.39$ m/s

Fig. 11 Velocity field in for the angles of 25°, 50° and 75°

Fig. 12 Comparison of 2D and 3D results in the context of vortex formation

4 Conclusions

The angle between the external carotid artery and the internal artery has the greatest impact on the following factors and phenomena: the thickness of the boundary layer, place of inflows of the concentration point and formation of vortexes (separation and flow retreat). The larger the areas of plaque deposition are, the wider the angle between the arteries. The following phenomena occur in the said zones:

- the velocity (kinetic energy) decreases;
- the turbulent Reynolds number and the kinetic energy of turbulence reach the maximum (at the cost of the decrease of the kinetic energy related to laminar flow);
- separation of the laminar layer and flow whirlpool occur—for angles wider than 60°;
- the viscosity grows (caused by decreased velocity);
- the impact of the concentration point on the velocity profiles at the outlets (thicknesses of the boundary layers) grows, because it causes acceleration of the fluid in

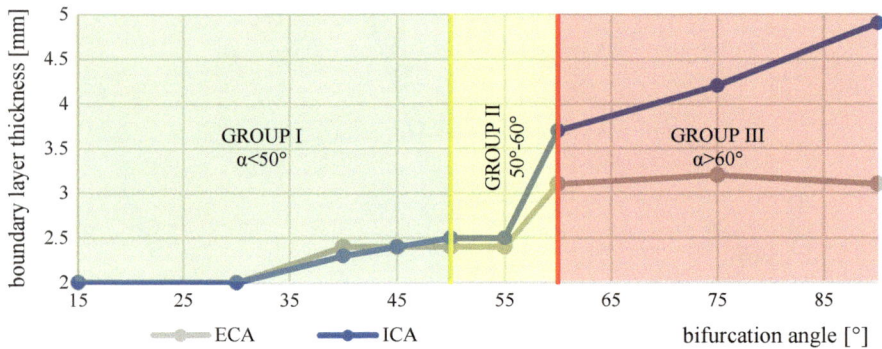

Fig. 13 Thickness of the boundary layer as a function of the bifurcation angle

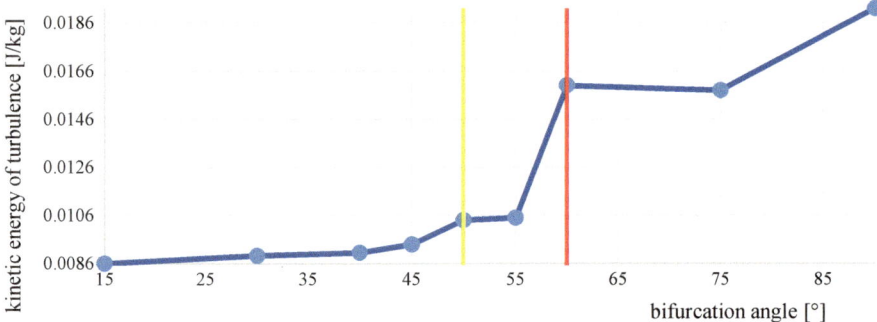

Fig. 14 Kinetic energy of turbulence as a function of the bifurcation angle

the areas behind the stagnation point. This area is additionally beneficial in terms of plaque deposition.

Based on the above prerequisites, the simulation results were divided into three principal groups. Figures 13, 14 and 15 present a series of diagrams reflecting the following relations: boundary layer thicknesses, kinetic energy of turbulence and turbulent Reynolds number, depending on the carotid artery bifurcation angle. A fast-growing trend of the above parameters for angles above 50° is particularly distinct.

GROUP I: angles below 50° → carotid arteries of geometries that cause no flow disorders, the change of value in the angle function is relatively constant.
GROUP II: angles in the range 50°–60° → carotid arteries, in which the parameters related to the flow turbulence grow, but occurring vortexes are very weak.
GROUP III: angles above 60° → carotid arteries of geometries enhancing the deposition of plaques with a tendency to separation and formation of vortexes.

Numerical results obtained in the 3D and 2D calculations were different. But the main tendency observed in the studies was preserved. Location of vortex formation and flow parameters values was very similar. The character of vortex formation was

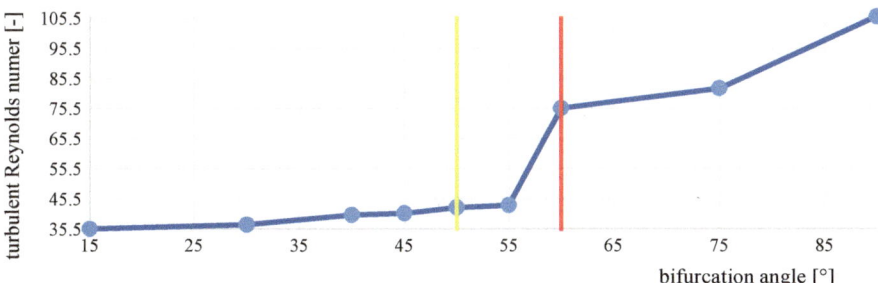

Fig. 15 Turbulent Reynolds number as a function of the bifurcation angle

different: in 3D geometry, vortices multiplied along the wall of a branched artery. Although the main purpose of the research has been achieved. It was possible to determine the location and the critical angle from which significant flow disturbances occur.

The results obtained in the studies confirm why patients with larger bifurcation angle are more susceptible to the occurrence of plaques in the arteries. The investigations disregarded a number of variables, such as incorrect blood work, patient's lifestyle, the thickness of the walls or the pulsating nature of the flow. These variables were disregarded because the authors focused on achieving an independent impact of bifurcation angle on the flow. Therefore, other parameters of the possible impact on the flow were omitted or simplified. The presented study results converge with the cases observed in practice—the largest group of patients with diagnosed deposits in the artery have the artery of the angle in excess of 50°. The points where deposits are formed very often occur near the bulb on the side of the internal artery, which is also confirmed by the performed studies. Based on the above prerequisites, a mechanically related confirmation was obtained as to why people of large carotid artery bifurcation angle are more exposed to stenosis.

The performed studies are a starting point for the development of a carotid artery geometry database planned in the future. The database will include the geometry groups that increase the probability of deposits. A person exposed to stenosis will have an opportunity to take proactive measures. Thanks to the combination of engineering practice related to flow modeling and medical practice related to cardiovascular surgery, biochemistry and biomechanics, numerous issues, yet to be solved, can be explained. The results may significantly contribute to the development of prophylactics and the improvement of the treatment of cardiovascular disease.

References

1. Van Steenhoven AA, Van de Vosse FN, Rindt CC, Janssen JD, Reneman RS (1990) Experimental and numerical analysis of carotid artery blood flow. Monogr Atheroscler, vol 15. Basel, Karger, pp. 250–260
2. Bijari PB, Wasserman BA, Steinman DA Carotid bifurcation geometry is an independent predictor of early wall thickening at the carotid bulb
3. Zarins CK, Giddens DP, Bharadvaj BK, Sottiurai VS, Mabon RF, Glagov S (1983) Carotid bifurcation atherosclerosis. Quantitative correlation of plaque localization with flow velocity profiles and wall shear stress. Circ Res 53:502–514
4. Wang HY, Liu LS, Cao HM, Li J, Deng RH, Fu Q, Zhang HX, Fei JG, Wang CX (2017) Hemodynamics in transplant renal artery stenosis and its alteration after stent implantation based on a patient-specific computational fluid dynamics model. Chin Med J (Engl) (2017)
5. Gharahi H, Zambrano BA, Zhu DC, Demarco JK, Baek S (2016) Computational fluid dynamic simulation of human carotid artery bifurcation based on anatomy and volumetric blood flow rate measured with magnetic resonance imaging HHS Public Access. Int J Adv Eng Sci Appl Math 8:40–60
6. Pedro Carvalho Rêgo de Serra Moura J, Presidente J, Lau Supervisor F, Manuel da Silva Chaves Ribeiro Pereira Co-supervisor J, Carlos Fernandes Pereira Vogais J, Bettencourt da Silva Pedro Álvares Serrão C (2011) Analysis and simulation of blood flow in the portal vein with uncertainty quantification aerospace engineering
7. Chung TJ (2009) Computational fluid dynamics
8. Schäfer M (2006) Computational engineering: introduction to numerical methods
9. Fluent AN (2009) 14.0 Tutorial guide. Ansys Inc
10. Moyle KR, Antiga L, Steinman DA (2006) Inlet conditions for image-based CFD models of the carotid bifurcation: is it reasonable to assume fully developed flow? J Biomech Eng 128:371
11. Moon JY, Suh DC, Lee YS, Kim YW, Lee JS (2014) Considerations of blood properties, outlet boundary conditions and energy loss approaches in computational fluid dynamics modeling. Neurointervention
12. Boyd J, Buick JM, Green S (2007) Analysis of the Casson and Carreau-Yasuda Non-Newtonian blood models in steady and oscillatory flows using the lattice Boltzmann method. Phys Fluids
13. Siebert MW, Fodor PS (2009) Newtonian and Non-Newtonian blood flow over a backward-facing step – a case study. In: Proceedings of the COMSOL conference 2009 Bost
14. Khan MF, Quadri ZA, Bhat SP (2013) Study of Newtonian and Non-Newtonian Effect of Blood Flow in Portal Vein in Normal and Hypertension Conditions using CFD Technique. Int J Eng Res Technol 6:974–3154
15. Chen J, Lu X-Y, Wang W (2006) Non-Newtonian effects of blood flow on hemodynamics in distal vascular graft anastomoses. J Biomech 39

Evaluating the Effect of Tissue Anisotropy on Brain Tumor Growth Using a Mechanically Coupled Reaction–Diffusion Model

Daniel Abler, Russell C. Rockne and Philippe Büchler

Abstract Glioblastoma (GBM) is the most frequent malignant brain tumor in adults and presents with different growth phenotypes. We use a mechanically coupled reaction–diffusion model to study the influence of structural brain tissue anisotropy on tumor growth. Tumors were seeded at multiple locations in a human MR-DTI brain atlas and their spatiotemporal evolution was simulated using the Finite Element Method. We evaluated the impact of tissue anisotropy on the model's ability to reproduce the aspherical shapes of real pathologies by comparing predicted lesions to publicly available GBM imaging data. The impact of tissue anisotropy on tumor shape was strongly location dependent and highest for tumors in brain regions with a single dominating white matter fiber direction, such as the corpus callosum. Despite strongly anisotropic growth assumptions, all simulated tumors remained more spherical than real lesions at the corresponding anatomic location and similar volume. These findings confirm previous simulation studies, suggesting that cell migration along WM fiber tracks is not a major determinant of tumor shape in the setting of reaction–diffusion-based tumor growth models and for most locations across the brain.

Keywords Glioma · Anisotropy · DTI · Mass effect · Reaction–diffusion model · Biomechanics

D. Abler (✉) · P. Büchler
ARTORG Center for Biomedical Engineering Research, University of Bern,
Bern, Switzerland
e-mail: daniel.abler@artorg.unibe.ch

P. Büchler
e-mail: philippe.buechler@artorg.unibe.ch

D. Abler · R. C. Rockne
Beckman Research Institute, City of Hope, Duarte, CA, USA
e-mail: rrockne@coh.org

© Springer Nature Switzerland AG 2019
J. M. R. S. Tavares and P. R. Fernandes (eds.), *New Developments on Computational Methods and Imaging in Biomechanics and Biomedical Engineering*, Lecture Notes in Computational Vision and Biomechanics 33, https://doi.org/10.1007/978-3-030-23073-9_3

1   Introduction

Gliomas are the most frequent primary brain tumors in adults (70%) [15]. Glioblastoma Multiforme (GBM) is the most malignant subtype of glioma, accounting for about 50% of diffuse gliomas. GBM infiltrates surrounding healthy tissue, grows rapidly, and forms a necrotic core of high cell density which is frequently accompanied by compression and displacement of the surrounding tissue. Despite aggressive treatment, long-term prognosis remains poor with median overall survival below 1.5 years [15].

Invasive growth and mass-effect are the macroscopic hallmarks of GBM. Variability can be observed with regard to these characteristics, ranging from predominantly invasive tumors without notable mass-effect to strongly displacing ones that induce higher mechanical stresses and result in healthy tissue deformation, midline shift, or herniation. These solid stresses play an important role for tumor evolution [9], which suggests that biomechanical factors have direct implications not only on the biophysical level, but may affect treatment response and outcome.

We have previously developed a mechanically coupled reaction–diffusion model of brain tumor growth that accounts for tumor mass-effect [1]. This framework simulates tumor evolution over time and across different brain regions using literature-based parameter estimates for tumor cell proliferation, as well as isotropic motility, and mechanical tissue properties. The model yielded realistic estimates of the mechanical impact of a growing tumor on intracranial pressure, however, comparison to imaging data showed that asymmetric shapes could not be reproduced.

To investigate the role of tissue anisotropy on simulated tumor shape, we extended our simulation framework to take into account tissue structure. White matter consists predominantly of aligned axonal fibers, whose orientation can be inferred from Magnetic Resonance (MR) Diffusion Tensor Imaging (DTI), which measures water diffusion along different directions in space. As diffusion is constrained transverse to fiber direction, MR-DTI provides structural information of brain tissue. Information from MR-DTI has previously been used to inform tumor cell migration behavior in mathematical models of brain tumor growth, see [17, table 1] for an overview of related work.

The few studies that have investigated the effect of tissue anisotropy on larger patient cohorts found it to have a beneficial, but relatively small effect on their models' ability to reproduce real tumor shapes. Employing the anisotropic glioma spread model of [14], [17] investigated the effect of tissue anisotropy without mass-effect. Their study on 10 cases showed an improved ability to approximate tumor shapes (average increase in Jaccard score by 0.03 ± 0.03, about 5% relative to the isotropic case) when including patient-specific DTI information and personalized estimates for a patient-specific anisotropy parameter that describes the sensitivity of cancer cells to the underlying brain structure. Only a few studies [3, 5, 6] took into account the tumor's mass effect when investigating the effect of tissue anisotropy. Simulation results of 9 low-grade glioma cases were reported by [6], using patient-specific DTI information, non-personalized growth parameters and an isotropic viscoelastic material model for brain tissue. Using DTI information in their study improved the Jaccard Score (Dice Index) between simulated and actual tumor by $\leq 2.40\%$ ($\leq 1.50\%$).

In the present study, we investigate the combined effect of anisotropic growth and mechanical tissue characteristics on tumor shape in a mechanically coupled reaction–diffusion model of invasive glioma growth by comparing simulation results obtained from isotropic and anisotropic material assumptions.

2 Materials and Methods

Figure 1 illustrates the study setup: Virtual tumors were seeded in an atlas of healthy brain anatomy at representative locations extracted from 10 subjects of the BRATS 2013[1] [11, 12] training dataset. Figure 2 shows the spatial distribution of the selected lesions in a human brain atlas. Tumor growth evolution was simulated for isotropic and anisotropic tissue properties and two sets of growth parameter choices, corresponding to diffuse and nodular growth characteristics, respectively. Virtually grown and real tumors were compared when the simulated tumor had reached the tumor volume of the corresponding subject from the BRATS dataset.

2.1 Mathematical Model

The mathematical model used in this study captures three interrelated aspects of macroscopic glioma growth [1]: Cell proliferation, invasion of tumor cells into the surrounding healthy tissue, and tissue deformation due to the tumor-induced mass-effect.

We model the invasive growth of glioma phenomenologically as a Reaction–Diffusion (RD) process [19], representing cell migration by passive diffusion:

$$\frac{\partial q}{\partial t} = \nabla \cdot \left(\hat{D} \nabla q \right) + \rho q \left(1 - q \right) , \tag{1}$$

with normalized cancer cell concentration $q(\mathbf{r}, t)$ and diffusion tensor $\hat{D} = \hat{D}(\mathbf{r})$. Tumor growth is modeled as a logistic growth process with proliferation rate ρ.

Similarly to [3, 5], the tissue-displacing mass-effect of the growing tumor is represented phenomenologically using a linear-elastic solid mechanics approach. It relies on the assumption that the creation of new tumor cells leads to volumetric

---

[1]https://www.smir.ch/BRATS/Start2013.

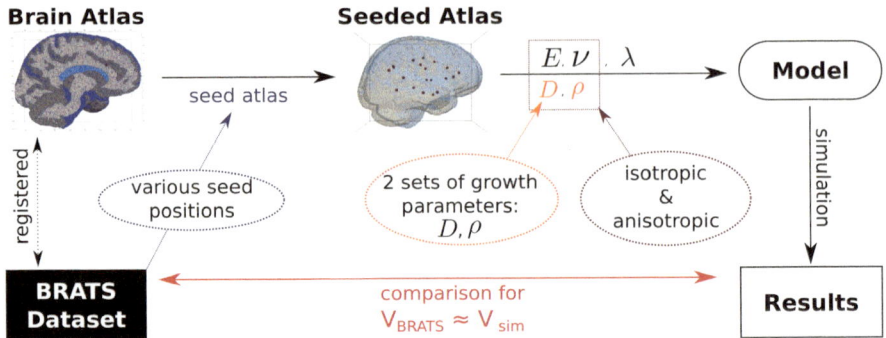

Fig. 1 Tumor growth evolution was simulated in a healthy brain atlas for two sets of growth parameters (D, ρ), and isotropic and anisotropic tissue properties. Simulated tumors were compared to subjects from the BRATS data set at approximately identical volume

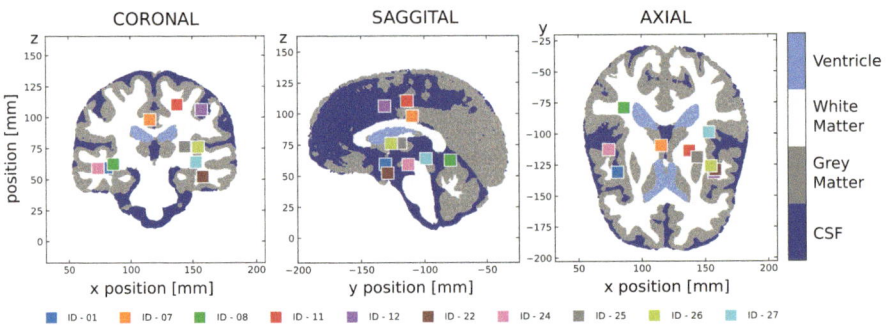

Fig. 2 Tumor center-of-mass positions of 10 selected BRATS cases projected onto central planes of SRI24 atlas

increase of the tumor and thus results in an expansion of the affected brain tissue. The volumetric increase is modeled by introducing a growth-induced strain component $\hat{\varepsilon}^{\text{growth}}(q)$, so that

$$\hat{\varepsilon}^{\text{total}}(\mathbf{u}, q) = \hat{\varepsilon}^{\text{elastic}}(\mathbf{u}) + \hat{\varepsilon}^{\text{growth}}(q) \ . \tag{2}$$

where displacements $\mathbf{u}$ are obtained from solving the linear-momentum equilibrium equation with stress $\hat{\sigma}(\mathbf{u})$ and strain $\hat{\varepsilon}^{\text{total}}(\mathbf{u})$ linked by a linear constitutive relationship.

Additionally, we assume a linear coupling between tumor cell concentration and growth-induced strain

$$\hat{\varepsilon}^{\text{growth}}(q) = \hat{\lambda} q = \lambda \mathbb{1} q \ , \tag{3}$$

with isotropic coupling strength λ.

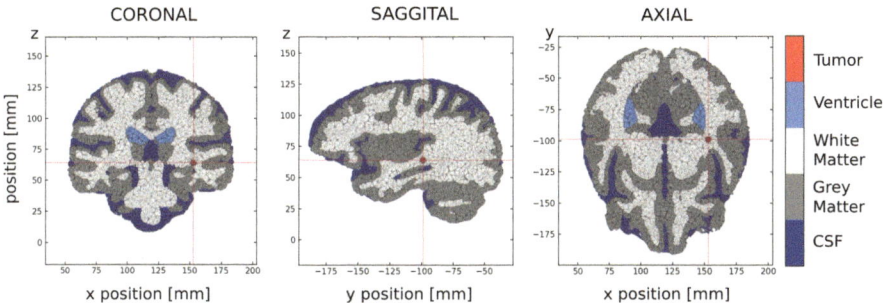

Fig. 3 Projections through seeded SRI24 atlas. An exemplary seed location is shown in the tetrahedral mesh used for simulation

2.2 Simulation Domain

We used the SRI24[2] [16] atlas of normal human brain anatomy to define the simulation geometry with tissue classes. Release 2.0 of the atlas provides separate tissue labels for White Matter (WM), Grey Matter (GM), and Cerebrospinal Fluid (CSF). We divided the CSF domain into two compartments to distinguish fluid-filled brain ventricles from the remaining CSF, surrounding the brain tissue. Additionally, the map of dominant Diffusion Tensor Imaging (DTI) eigenvectors was obtained from an earlier release (v0.0) of the atlas. This information was interpreted as local orientation of axon fibers and was used to inform diffusion and mechanical tissue parameters in the anisotropic simulation scenario. Finally, all relevant components of the atlas were registered to fit the spatial orientation of the BRATS datasets.

The tumor center-of-mass position was computed for each of the 10 selected subjects, based on the tumor volume visible on T1-weighted contrast-enhanced (T1c) MR imaging. For each subject, a spherical tumor seed (2 mm radius) was introduced in the atlas label map at the corresponding center-of-mass position, and a tetrahedral mesh was generated (approximately 320 000 elements) using CGAL[3] and VTK[4] libraries. DTI information from the SRI24 atlas was then interpolated over the seeded mesh. Figure 3 shows coronal, sagittal, and axial views through an exemplary seeded and meshed simulation domain.

---

[2]https://www.nitrc.org/projects/sri24/.

[3]https://www.cgal.org.

[4]https://www.vtk.org.

2.3   Simulation Assumptions

To compare the effect of tissue anisotropy on the evolution of tumor characteristics, two different simulation scenarios were considered, assuming isotropic and
anisotropic material properties, respectively.

In both cases, the brain tissues WM and GM were modeled as linear-elastic
materials. The CSF of the brain ventricles was modeled as compressible to account
for physiological mechanisms that compensate elevated intracranial pressure [21],
whereas the remaining CSF was modeled as nearly incompressible. Simulations
were run with two distinct sets of growth parameters corresponding to *nodular* and
diffuse growth characteristics with $\rho/D \geq 1.35\,\text{mm}^{-2}$ and $\rho/D \leq 0.37\,\text{mm}^{-2}$ [2],
respectively. A maximum tumor-induced strain of 15% [13] was assumed, $\lambda = 0.15$,
and an initial condition of $q = 1$ over the entire volume of the tumor seed was
imposed. Deformation of the brain surface and escape of tumor cells from the brain
were constrained by zero-displacement and zero-flux boundary conditions at surface
nodes. The mathematical model was solved using the Finite Element Method. It
was implemented in Abaqus (Simulia, Dassault Systémes) as fully coupled thermal
stress analysis using 4-node linear elements (C3D4T) with the tumor mass-effect
being represented by volumetric thermal expansion.

Isotropic Scenario

In the isotropic simulation scenario, diffusion and mechanical tissue behavior were
assumed isotropic using the parameter values summarized in Tables 1 and 2 for the
considered tissue types. The linear material model was fully characterized by two

Table 1 Reaction–diffusion parameter sets (D, ρ), representing *nodular* and *diffuse growth*. Tissue-
specific motility estimates (D_{WM}, D_{GM}) are based on the assumption that D_{avg} was measured in a
tissue volume containing equal portions of GM and WM, and $D_{WM} = 5 D_{GM}$ [19]

Growth type	ρ [1/d]	D_{avg} [mm²/d]	D_{avg}/ρ [mm²]	ρ/D_{avg} [mm⁻²]	D_{GM} [mm²/d]	D_{WM} [mm²/d]
Nodular	0.082	0.053	0.650	1.540	0.020	0.101
Diffuse	0.037	0.105	2.855	0.350	0.040	0.200

Table 2 Mechanical tissue properties (isotropic case), informed by [21]

Tissue	E [kPa]	ν
W/G Matter	3.0	0.45
Tumor	6.0	0.45
CSF (Ventricles)	1.0	0.30
CSF (other)	1.0	0.49

parameters, Poisson ratio ν and Young's modulus E. Values for Young's modulus of brain tissue and tumor were adopted from [21].

Anisotropic Scenario

In the anisotropic simulation scenario, white matter fiber directionality was taken into account and the tissue was modeled as transversely isotropic material with different material properties along ($\parallel$) and orthogonal to ($\perp$) the fibers.

Tumor cell motility along fiber direction (WM) was assumed identical to the isotropic case $D_{\mathrm{W}}^{\parallel} = D_{\mathrm{W}}^{\mathrm{iso}}$, whereas a significantly lower motility was chosen for the transverse direction $D_{\mathrm{W}}^{\perp} = 0.01 D_{\mathrm{W}}^{\mathrm{iso}}$. Due to reduced fiber alignment, cell motility in grey matter was modeled as isotropic [3, 5] with the value indicated in Table 1. We chose a very high ratio $D_{\mathrm{W}}^{\parallel}/D_{\mathrm{W}}^{\perp} = 100$ to investigate the effect of growth anisotropy. For comparison, [6] assumed a ratio of 5; [10] varied this ratio between 5 and 100 and found the best "de visu" fit for a ratio of 10.

Linear-elastic mechanical tissue properties of the transversely isotropic situation can be expressed in terms of seven engineering constants: Two Young's moduli that describe the stresses resulting from uniaxial stretch parallel $E^{\parallel}$ and perpendicular $E^{\perp}$ to the fiber axis. Two shear moduli that describe shear stresses in planes parallel to ($\mu^{\parallel}$) and normal to ($\mu^{\perp}$) the fiber axis. Three Poisson ratios $\nu^{\parallel\perp}$, $\nu^{\perp\parallel}$, $\nu^{\perp\perp}$ that describe the strain in one direction ($\parallel$ or $\perp$) that arises from stretch in another orthogonal direction ($\parallel$ or $\perp$). Only five of these seven parameters are typically independent since additionally:

$$\frac{\nu^{\parallel\perp}}{E^{\parallel}} = \frac{\nu^{\perp\parallel}}{E^{\perp}} \tag{4a}$$

$$\mu^{\perp} = \frac{E^{\perp}}{2(1+\nu^{\perp\perp})} . \tag{4b}$$

To estimate parameters of that model, we assume a fiber reinforcement effect in white matter that increases resistance against stretch along the fiber direction, $E_{W}^{\parallel} = 3 \cdot E_{W}^{\perp}$, from observations on lamb corpus callosum $E^{\parallel}/E^{\perp} \approx 6.5$ [7] and porcine corona radiata $E^{\parallel}/E^{\perp} \approx 2.7$ [7, 20]. Based on the material parameters used for the isotropic cases, we defined the Young's moduli of white matter so that $E_{\mathrm{WM}}^{\parallel} > E_{\mathrm{GM}}^{\mathrm{iso}} > E_{\mathrm{WM}}^{\perp}$. We assume $\nu^{\perp\parallel} = \nu^{\mathrm{iso}}$, so that $\nu^{\perp\parallel}$ follows from Eq. (4a) and $\nu^{\perp\perp} = 1 - \nu^{\perp\parallel}$. This allows us to compute $\mu^{\perp}$ from Eq. (4b). We then compute $\mu^{\parallel} = 1.4\,\mu^{\perp}$ [7]. Resulting mechanical model parameters for white matter are summarized in Table 3.

2.4 Analysis

Two different tumor detection thresholds were used to evaluate simulation results: $c_{\mathrm{T1c}} = 0.80$ and $c_{\mathrm{T2}} = 0.16$ corresponding to tumor features visible on T1-weighted contrast enhanced (T1c) and T2-weighted (T2) MRI imaging [18], respectively. Simulations were run until the simulated tumor had reached the T1c volume of the corresponding BRATS subject. Tumors corresponding to T1c and T2 visibility threshold were extracted at multiples of 5mm increments in equivalent radius computed from the simulated T1c volume. For each tumor volume, the following measures were computed: (a) Tumor *aspect ratio*, as the ratio between the shortest and longest axis of the smallest bounding box around the tumor. A value of 1 corresponds to a spherical tumor shape; values closer to 0 indicate aspherical (elongated, oblate, or asymmetric) shapes. (b) Tumor *nodularity*, as the ratio of T1c and T2 tumor volumes. A value close to 1 corresponds to a very well delineated, nodal tumor, whereas values closer to 0 indicate diffuse growth.

The same measures were computed from BRATS segmentations by identifying the T1c tumor volume with labels {necrotic, non-enhancing tumor, enhancing tumor} and the T2 volume with labels {necrotic, non-enhancing tumor, enhancing tumor, edema}. Measures derived from simulated tumors and real pathologies were compared at similar volumes $V_{\text{T1c, sim}} \approx V_{\text{T1, BRATS}}$.

3 Results

Tumor growth evolution and tissue deformation were simulated for all 10 selected BRATS subjects, growth parameterizations (nodular, diffuse) and tissue structure scenarios (isotropic, anisotropic).

The anisotropic growth scenario showed an average 4.3 ± 6.2 % reduction of tumor aspect ratio compared to isotropic growth assumptions. The impact on tumor shape was similar for *diffuse* $(3.9 \pm 7.6\%)$ and *nodular* $(4.7 \pm 4.6\%)$ growth parameterizations. However, both isotropic and anisotropic growth assumptions resulted in simulated tumor shapes that were more spherical than the corresponding BRATS lesions, Fig. 4.

The effect of tissue anisotropy on simulated tumor shape was strongly dependent on seed location: Tumors grown from seeds located deep in WM (ID-07, ID-27) and adjacent to the lateral ventricle (ID-08) exhibited a strong effect of tissue anisotropy.

Table 3 Mechanical tissue properties (anisotropic simulation scenario), assuming transverse symmetry with directions along (∥) and orthogonal to (⊥) fiber direction. Material properties for GM and CSF were those from Table 2

Tissue	$E^{\parallel}$ [kPa]	$E^{\perp}$ [kPa]	$\mu^{\parallel}$ [kPa]	$\mu^{\perp}$ [kPa]	$\nu^{\parallel\perp}$	$\nu^{\perp\parallel}$	$\nu^{\perp\perp}$
White matter	4.5	1.5	0.56	0.40	0.45	0.15	0.85
Tumor (if in WM)	9.0	3.0	1.12	0.8	0.45	0.15	0.85

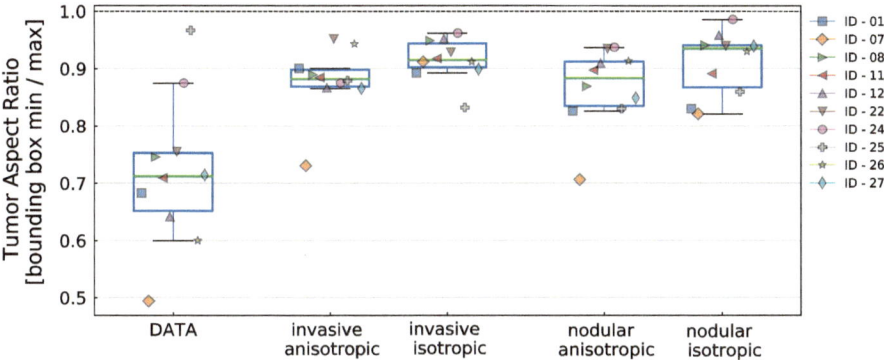

Fig. 4 Aspect ratio of BRATS T1c lesions and simulated tumors for diffuse/nodular growth parameterization and isotropic/anisotropic tissue properties. A value of 1 indicates a spherical shape, whereas lower values correspond to oblate or elongated shapes

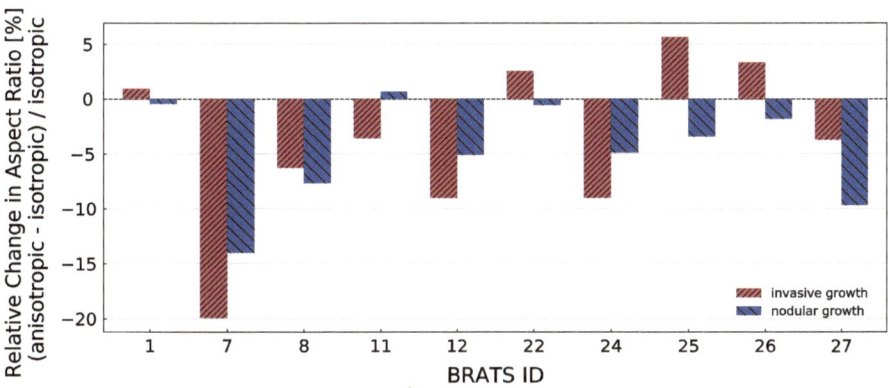

Fig. 5 Relative change in tumor aspect ratio between isotropic and anisotropic configurations. A negative value corresponds to a decrease in aspect ratio due to anisotropic material properties

Seeds located closer to WM/GM interfaces (ID-11, ID-12, ID-22, ID-24) showed mixed effects, while those located in GM (ID-01, ID-25, ID-26) experienced only small effects, Fig. 5. These observations are consistent with our parameterization which considers GM to be isotropic. The effect of tissue anisotropy on shape was particularly pronounced for ID-07 located medially in the corpus callosum, a region of highly aligned axons.

Tumor nodularity extracted from BRATS images (*DATA* in Fig. 6) differed across the selected cases. For each simulated BRATS case, the computed nodularity measure was consistent with growth parameterization: lower for diffuse and higher for nodal growth. In most cases, the anisotropic growth scenario resulted in more nodular tumors compared to isotropic growth assumptions, due to reduced overall diffusivity. Despite identical growth parameterization (nodular, diffuse), the computed nodularity of simulated tumors differed across BRATS subjects. This effect can be

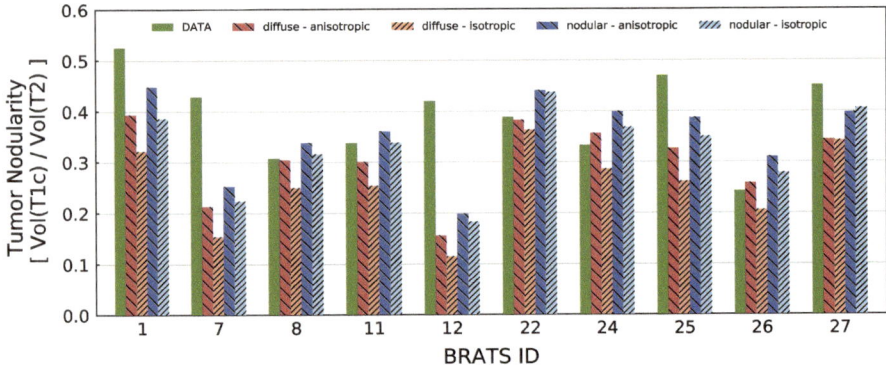

Fig. 6 Tumor nodularity of BRATS lesions and simulated tumors for diffuse/nodular growth parameterization and isotropic/anisotropic tissue properties. A value close to 1 corresponds to a nodular tumor, whereas smaller values indicate diffuse growth

attributed to differences in the growth environment (WM, GM, boundary, constrained by CSF/ventricle) resulting in distinct average growth parameters for each lesion.

4 Discussion

This study explored the effect of tissue anisotropy on glioma growth simulations in a 3D human brain atlas. In agreement with model parameterization, we found tissue anisotropy to result in reduced tumor shape symmetry for tumors located in WM and for some lesions at the WM/GM interface. However, despite choosing strongly anisotropic diffusion parameters, $D_W^{\parallel}/D_W^{\perp} = 100$, all simulated tumors remained more spherical than real lesions at the corresponding anatomical location and of similar volume.

Our findings confirm findings of previous simulation studies [6, 17] suggesting that anisotropic cell migration along WM fiber tracks is not a major determinant of tumor shape in the setting of reaction–diffusion-based tumor growth models and for most locations across the brain. Exceptions might apply for tumors located in brain regions where a single dominant fiber direction prevails throughout a larger contiguous volume segment. For example, in this study, we observed the highest relative change in aspect ratio due to tissue anisotropy, 14–20%, for a medially located GBM in the corpus callosum (ID-07).

Large variability in tumor nodularity for identical growth parameterizations (diffuse/nodular) across different brain locations, Fig. 6, indicates that 3D tumor growth is strongly affected by the tissue composition of a tumor's growth domain. We hypothesize that the interplay between tissue composition, spatial constraints, and resulting mechanical forces may exceed the effect of tissue anisotropy on tumor growth, possibly giving rise to location-specific growth archetypes of GBM.

While our model computed tumor mass-effect and resulting in healthy tissue deformation, neither this nor similar previous modeling studies for human GBM [6, 17] captured the growth-inhibiting effect of solid stress [8].

The present study has further limitations: (a) Only the tumor seed position was personalized for each growth model, not the brain anatomy or growth parameters. A mismatch between patient and atlas anatomy and/or asymmetric growth may have resulted in the simulated tumor growth process being initialized in a different brain tissue, which can significantly affect the tumor's simulated evolution. This may explain shape discrepancies for some of the BRATS cases, such as ID-25 (ID-26) which has a very high (low) aspect ratio in the BRATS dataset, but ranges among the simulated tumor shapes with lowest (highest) aspect ratio. (b) DTI information was derived from an atlas of the healthy human brain so that possible changes in local tissue structure due to tumor growth could not be taken into account. We considered brain tissues to be either isotropic (GM) or anisotropic (WM), not distinguishing varying degrees of anisotropy within each tissue class. Also, possible differences in patient-specific sensitivity of cancer cells to the underlying brain structure were not taken into account. (c) This study relied on a linear-elastic material model with estimates for mechanical tissue anisotropy derived from animal brain tissue characterization. Recent evidence [4] suggests that an Ogden material model captures the mechanical response of human brain tissue more accurately.

5 Conclusion

This study investigated the joint impact of tumor mass effect and tissue anisotropy on simulated tumor shape. In agreement with previous simulation studies, we find that anisotropic cell migration along WM fiber tracks is not a major determinant of tumor shape, except for growth locations where a single dominant fiber direction prevails throughout a larger contiguous volume segment. Further work is needed to combine the individual contributions of structural anisotropy, tissue composition, and mechanical growth constraints in a way to best reproduce GBM growth characteristics.

Acknowledgements The research leading to these results has received funding from the European Union's Horizon 2020 research and innovation programme under the Marie Skodowska-Curie grant agreement No 753878. Calculations were performed on UBELIX (http://www.id.unibe.ch/hpc), the HPC cluster at the University of Bern.

References

1. Abler D et al (2018) Evaluation of a mechanically coupled reaction-diffusion model for macroscopic brain tumor growth. In: Gefen A et al (eds) Computer methods in biomechanics and biomedical engineering. Springer International Publishing, Cham, pp 57–64

2. Baldock AL et al (2014) Patient-specific metrics of invasiveness reveal significant prognostic benefit of resection in a predictable subset of gliomas. LoS ONE 9(10):e99057
3. Bondiau P-Y et al (2008) Biocomputing: numerical simulation of glioblastoma growth using diffusion tensor imaging. Phys Med Biol 53(4):879–893
4. Budday S et al (2017) Mechanical characterization of human brain tissue. Acta Biomater 48:319-340
5. Clatz O et al (2005) Realistic simulation of the 3-D growth of brain tumors in MR images coupling diffusion with biomechanical deformation. IEEE Trans Med Imaging 24(10):1334–1346
6. Elazab A et al (2017) Post-surgery glioma growth modeling from magnetic resonance images for patients with treatment. Sci Rep 7(1)
7. Feng Y et al (2013) Measurements of mechanical anisotropy in brain tissue and implications for transversely isotropic material models of white matter. J Mech Behav Biomed Mater 23:117–132
8. Helmlinger G et al (1997) Solid stress inhibits the growth of multicellular tumor spheroids. Nat Biotechnol 15(8):778–783
9. Jain RK et al (2014) The role of mechanical forces in tumor growth and therapy. Annu Rev Biomed Eng 16(1):321–346
10. Jbabdi S et al (2005) Simulation of anisotropic growth of low-grade gliomas using diffusion tensor imaging. Magn Reson Med 54(3):616–624
11. Kistler M et al (2013) The virtual skeleton database: an open access repository for biomedical research and collaboration. J Med Internet Res 15(11):e245
12. Menze B et al (2014) The multimodal brain tumor image segmentation benchmark (BRATS). IEEE Trans Med Imaging, p 33
13. Mohamed A et al (2005) Finite element modeling of brain tumor mass-effect from 3D medical images. In: Medical image computing and computer-assisted intervention-MICCAI 2005. Lecture notes in computer science 3749. Springer, Berlin, pp 400–408
14. Painter K et al (2013) Mathematical modelling of glioma growth: the use of diffusion tensor imaging (DTI) data to predict the anisotropic pathways of cancer invasion. J Theor Biol 323:25–39
15. Ricard D et al (2012) Primary brain tumours in adults. Lancet 379(9830):1984–1996
16. Rohlfing T et al (2009) The SRI24 multichannel atlas of normal adult human brain structure. Hum Brain Mapp 31(5):798–819
17. Swan A et al (2017) A patient-specific anisotropic diffusion model for brain tumour spread. Bull Math Biol
18. Swanson KR et al (2008) A mathematical modelling tool for predicting survival of individual patients following resection of glioblastoma: a proof of principle. Br J Cancer 98(1):113–119
19. Swanson KR et al (2000) A quantitative model for differential motility of gliomas in grey and white matter. Cell Prolif 33(5):317–329
20. Velardi F et al (2006) Anisotropic constitutive equations and experimental tensile behavior of brain tissue. Biomech Model Mechanobiol 5(1):53–61
21. Wittek A et al (2010) Patient-specific non-linear finite element modelling for predicting soft organ deformation in real-time; application to non-rigid neuroimage registration. Prog Biophys Mol Biol. Special Issue on Biomechanical Modelling of Soft Tissue Motion 103(2–3):292–303

Prediction of Stress and Strain Patterns from Load Rearrangement in Human Osteoarthritic Femur Head: Finite Element Study with the Integration of Muscular Forces and Friction Contact

Fabiano Bini, Andrada Pica, Andrea Marinozzi and Franco Marinozzi

Abstract Osteoarthritis (OA) is a degenerative disease that alters the integrity of the joint. Osteophytes represent abnormal osteocartilaginous outgrowths associated with the evolution of OA. Finite element (FE) analysis was performed on an 3D model of the proximal half of human femur to determine the relevance of osteophytes on the stress and strain distributions within the femur head. We assume that the model includes three linearly elastic, homogeneous and isotropic media representing the articular cartilage, the cortical and trabecular bone. With the aim of a more accurate representation of the physiological conditions, we consider in the FE model the influence of the muscle forces that span the hip joint. We also assume a friction contact between the cartilage layer and the cortical tissue. Simulations were carried out for a healthy and three different stages of OA femur. Different load distributions are considered for the four models due to the alterations of bone structure. The patterns of stress and strain within the trabecular tissue suggest that osteophytes manifestation could justify the development of bone cysts (geodes) and the formation of highly mineralized tissue (eburnation). The FE approach presented in this work could result useful in predicting bone behaviour towards abnormal mechanical solicitations.

1 Introduction

Osteoarthritis (OA) is a disease that alters the integrity of the tissues that compose the joint and causes a significant loss of mobility, mainly among the adult population. It represents a significant and expensive issue in the public health. In fact, with the increase of ageing population, OA is expected to become the fourth leading cause

F. Bini (✉) · A. Pica · F. Marinozzi
Department of Mechanical and Aerospace Engineering, Sapienza University of Rome, via Eudossiana, 18, 00184 Rome, Italy
e-mail: fabiano.bini@uniroma1.it

A. Marinozzi
Orthopedy and Traumatology Area, Campus Bio-Medico University, via Alvaro del Portillo, 200, 00128 Rome, Italy

© Springer Nature Switzerland AG 2019
J. M. R. S. Tavares and P. R. Fernandes (eds.), *New Developments on Computational Methods and Imaging in Biomechanics and Biomedical Engineering*, Lecture Notes in Computational Vision and Biomechanics 33, https://doi.org/10.1007/978-3-030-23073-9_4

of disability by 2020 [1]. Therefore, it emerges the crucial importance to better understand the disease progression and to predict the structural consequences of the joint degradation.

Generally, OA is characterized by progressive cartilage degeneration, alteration of subchondral bone structure, osteophytes formation and synovial fibrosis [2]. Although it is often described as an articular cartilage disease, new evidence highlights that changes of the structural and material properties of bone are not secondary manifestations of OA, but active contributors to the disease progression. Altered load distributions can accelerate the evolution of OA [3] and lead to the aberrant remodelling processes of the joint functional units.

The importance of bone alterations in OA progression is still not properly understood. During the OA development, detectable alterations in the composition and structure of bone tissue appear prior to the cartilage degeneration [4]. Therefore, it is worth investigating the bone variations in order to achieve an early identification of the disease. The OA alterations in bone include increases of subchondral cortical bone thickness, decreases in subchondral trabecular bone mass, formation of osteophytes and cysts [4]. Osteophytes are osteocartilaginous outgrowths that promote the creation of contact points between the bony extremes of osteoarthritic joints and participate in the development of mechanical forces alterations within the diseased articulations [5–7]. Osteophytes are developed by a process of endochondral ossification, already observed in the primary osteogenesis [4]. This mechanism of bone formation is characterized by a rapid and unorganized mineralization of the cartilaginous matrix [8].

Osteophytes develop at sites of tendon insertion or they arise from the periosteum covering the bone [9]. Radiographic histological study [5] has identified different distributions of osteophytes, i.e. in the peripheral zone of the femoral head (marginal osteophytes), on the medial surface of the femoral head (epiarticular osteophytes) and across the basal layers of the articular cartilage (subarticular osteophytes). In this study, depending on the OA severity, we analysed the scenarios of marginal, epiarticular and subarticular osteophytes developed on the human femur head.

Previous studies [10, 11] based on an 2D finite element model of the femur head, provided insights into the alterations of the strain and stress distribution subsequent to the development of osteophytes. In the current study, we aim to simulate more accurate loading conditions for the proximal femur by taking into account the influence of an ensemble of muscular forces that acts on the great trochanter. We develop the present analysis assuming that the load associated with the stance phase of gait is transferred in friction conditions between the cartilage layer and the femur head. The stress and strain distributions achieved from a healthy femur were compared to the outcomes of three pathological models characterized by different altered load distributions on the femur head in order to mimic the presence of osteophytes. The overall structural variations predicted in the trabecular region are in agreement with clinical observations [5, 12, 13] and previous FEM studies [14–16].

2   Methods

An 3D model of the proximal half of the femur head was implemented from Computer Tomography images. The image slices of the femur head of a 72-year-old male were provided by the National Library of Medicine in Maryland (USA). In the 3D geometric model, the centre of the Cartesian coordinate system coincides with the centre of the femur head. The femoral Z-axis is vertical, the Y-axis points in the medial-lateral direction and the X-axis points in the posterior–anterior direction. The Y- and Z-axes are in the coronal plane of the body, while the X-axis is in the sagittal plane.

The 3D FE model is characterized as a multilayer solid, composed of an inner trabecular region, an external cortical zone and a cartilage shell which covers the femoral head [17, 18]. In the present study, cartilage thickness is maintained constantly on the whole surface of the femoral head. Since it was observed that osteophytes formation can occur prior to the degeneration of the overlying cartilage [19], we also assume that the cartilage thickness remains unaltered during OA evolution. For the accurate estimation of the mechanical response of the femur head model, the geometry was meshed with tetrahedral elements with an averaged minimum size of 0.1 mm. In Table 1, we provide the FEM statistics of the FE simulations performed.

Tissue properties were modelled to be constant, linearly elastic, isotropic and homogeneous. All material properties were chosen in accordance with related literature. The trabecular tissue is characterized by a Young's modulus of 1 GPa [20–22], a Poisson's ratio of 0.3 [23] and an apparent tissue density of 1000 kg/m^3 [24]. For the cortical bone, a Young's modulus of 22 GPa [25], a Poisson's ratio of 0.3 [26] and a tissue density equal to 2000 kg/m^3 [27] is assigned. The articular cartilage is assumed to have an average value of the Young's modulus equal to 15 MPa [28], a Poisson's ratio of 0.1 [29] and a density of 1000 kg/m^3 [3]. The above numerical considerations remain unchanged for both healthy and pathological models. Appropriate boundary conditions allow to highlight the evolution of the disease.

FEM simulations are performed for a healthy femur (HF) and for three different levels of OA severity, i.e. early stage (ES), intermediate stage (IS) and advanced stage (AS). The early stage condition is characterized by the presence of two groups of marginal osteophytes (O_1 and O_2). In the IS model, a foveal group of osteophytes (O_3) is added to the existing peripheral groups. The AS case is represented by four

Table 1 Statistics of FE simulations

Model	Degrees of freedom	Number of mesh elements	Minimum mesh element (mm)
HF	3342539	615995	0.1
ES	2550538	471405	0.1
IS	2830439	534242	0.1
AS	3223518	604683	0.1

Fig. 1 3D model of proximal half of human femur. The red arrow indicates the resultant muscular force due to the presence of three muscles. The arrow also illustrates the attachment location

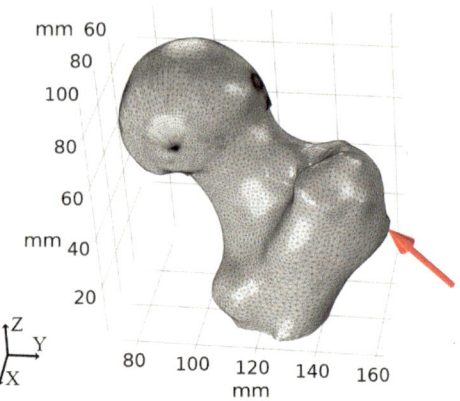

groups of osteophytes, namely two marginal osteophytes, an epiarticular (O_3) and a subarticular (O_4) bony outgrowths.

In order to represent the evolution of the disease in the FE model, the variations of geometrical parameters concerning the acetabular coverage of the femoral head and different load distributions are taken into account. The load is transferred between the acetabulum and the femur head by means of the contact surface defined by three centre-edge angles: the centre-edge angle of Wiberg in the YZ plane, the anterior ($\theta_{A\text{-}CE}$) and posterior ($\theta_{P\text{-}CE}$) centre-edge angles in the sagittal plane, i.e. XZ plane [30] see Fig. 1 in [14]. The extension of the main contact area (MCA) in the YZ plane is identified by the functional angle θ_F. In the healthy femur condition, the centre-edge angle of Wiberg (θ_{CE}) in the YZ plane is assumed to be equal to 30°, while the angle θ_F is set to 110°.

Generally, the pathological cases are characterized by values of θ_{CE} minor than 20° [31], thus we adopted a value of 10° in the OA models. In the coronal plane, we assume that the presence of osteophytes acts in detriment of the extension of the MCA. Thus, we make the hypothesis that in the ES model the functional angle θ_F is equal to 50°, in the IS model θ_F is set to 40° while in the AS case θ_F is 35°.

Conversely, the surface occupied by each group of osteophytes increases with the degeneration of the disease. We make the assumption that the osteophytes characterizing an OA stage have the same extension, defined by the angle θ_{Oi}. Namely, the ES model is defined by the angle θ_{Oi} equal to 15°, in the IS model θ_{Oi} is set to 20°, while in the AS model θ_{Oi} is 25°. In the sagittal plane, the extension of the MCA is maintained constant, both in healthy and pathological conditions. Furthermore, we assume that the groups of osteophytes have the same extension as the MCA. According to clinical observation [30], we set both $\theta_{A\text{-}CE}$ and $\theta_{P\text{-}CE}$ to 60°.

We considered that the musculoskeletal loading conditions at the hip are determined by the joint contact force and the forces of the muscles that span the hip joint. In the single-legged stance phase of gait, the loading force (H) acting on the hip joint is determined by the partial body mass (W), obtained as total body mass diminished by the weight-bearing leg, and the abductor muscle force calculated to be two times

W [32]. The magnitude of H is approximatively 2.4 times W, in agreement with [33]. For a subject with a body mass estimated to 65 kg, the loading force H corresponds to 1554 N.

The resultant force H acting between the acetabulum and the femur head was directed normally to the contact surface of the femur head. In the healthy femur, the total force H acts on the contact surface between the acetabulum and the femur head, i.e. MCA. In pathological conditions, the force is distributed among various contact regions, namely the reduced MCA and the different groups of osteophytes. In accordance with [10, 11], in the ES, IS and AS models, the load acting on the MCA is, respectively, 75, 50 and 25% of the loading force H. The remaining percentages were equally distributed between the two, three and four groups of osteophytes which characterize the OA level of severity.

Normal and pathological biomechanical conditions were simulated implementing different loading patterns in order to obtain contact pressure distribution during the stance phase of gait. We applied the load profile as a boundary condition on each domain that composes the contact surface in the FE model of the femur head. In the coronal plane, we divided the contact surface into circular sectors characterized by a centre-edge angle of 5°. Globally, the contact pressure (p) integrated over the articular contact surface (A) is equal to the force H transmitted to the hip joint, as indicated by the following integral relationship [34]:

$$\int p \cdot dA = H \tag{1}$$

The real hip joint is characterized as a ball and socket configuration. In order to model the transmission of force across the hip joint, we consider different pressure patterns in the healthy and OA conditions [35, 36]. In the healthy femur condition, the pressure distribution within the hip joint in the YZ plane could be described by a cosine distribution, as indicated by Eq. (2) [34, 37]:

$$p = p_{max} \cdot \cos \gamma \tag{2}$$

where p_{max} is the maximum contact pressure and γ is the angle between a generic point on the MCA and the Z-axis, See Fig. 1 in [14].

The maximum pressure is calculated as follows:

$$p_{max} = \frac{H}{\sum_{i=1}^{n} \cos(\gamma_i)} \cdot \frac{1}{A} \tag{3}$$

where γ_i is the value of the angle γ at the distal extremity of each domain of the MCA from the Z-axis, n is the number of the domains and A is the area of the contact surface.

In pathological cases, concentrated loads are supposed to be experimented by the femoral head surface because of the osteophytes appearance and the reduction of the MCA. In reviewing the literature associated with contact interaction [36, 38, 39], the pressure pattern due to the load transfer on the femur head via the thin elastic layer of the cartilage is modelized with a symmetric parabolic distribution. According to Johnson [36], the pressure distribution can be described as a function of the distance **r** from the centre of the contact area of radius **a** as expressed by Eq. (4) [36, 38], where p_{max} is the maximum contact pressure:

$$p(r) = p_{max} \cdot \left[1 - \left(\frac{r}{a}\right)^2\right]^{\frac{1}{2}} \tag{4}$$

The maximum contact pressure is expressed as

$$p_{max} = \frac{k \cdot H}{A} \tag{5}$$

where the coefficient k indicates the percentage of the total force H which acts on the contact area A. The latter can be referred to as the reduced MCA or to the contact surface of the osteophytes.

The pressure distributions acting on the reduced MCA, on the foveal, epiarticular and subarticular osteophytes are described by the axisymmetrical parabolic profile described previously. The maximum pressure value occurs in the centre of the contact area and null in the peripheral zone, at the maximum distance from the centre.

The marginal osteophytes are characterized as half-spheres impinging on the acetabular labrum. In order to respect the acetabular coverage conditions, we considered a half parabolic distribution in the YZ plane. Therefore, the maximum pressure is obtained in correspondence of the acetabular rim in agreement with [40] that indicate this region as one of the most solicited of the femur head.

In the sagittal plane, for both healthy and pathological models, we adopt a symmetric parabolic distribution of the pressure on the femur head. The maximum value is achieved in the centre of the contact surface, which is assumed to be in correspondence of the Z-axis [39], see Fig. 1 in [14].

The major muscles which act on the proximal human femur during the stance phase of gait are included in the model, i.e. gluteus minimus, gluteus medius, gluteus maximus and the tensor fascia latae proximal and distal parts [15]. As a simplification, muscle forces were considered by grouping the muscles with similar functions. Therefore, the gluteal muscles were combined into a single force vector. For each force, an insertion point is defined at the great trochanter. In this study, the three forces were assumed to be applied at the same point [41] (Fig. 1). Similarly as in other studies [42, 43], the magnitude and direction of the muscle forces were taken from literature, namely from the study of Heller et al. [41], based on the investigations of Duda et al. [15] and Brand et al. [44]. (Table 2) The magnitude of the forces corresponding to the stance phase of gait was scaled to the bodyweight, following the approach presented in [15]. We assume that the muscle forces remain unchanged during the evolution of the disease.

A further refinement in the modelling approach consists in taking into account the contribution of the friction contact between the cartilage and the cortical tissue.

Table 2 The forces (in percentage of bodyweight) and the coordinates (in millimetres) of the insertion point are given in the coordinate system of the femur

Muscle name	Force magnitude (% BW)			Acts at point
	X	Y	Z	
Gluteal muscles	4.3	−58	86.5	P1 (12.04, 67.83, −35.45)
Tensor fascia latae proximal part	11.6	−7.2	13.2	
Tensor fascia latae distal part	−0.7	0.5	−19	

In the FE analysis, the frictional force is considered proportional to the normal force acting on the femoral head with a static coefficient of friction, μ, as in Eq. 6:

$$F_t = \mu \cdot H \tag{6}$$

In agreement with experimental studies [45], we adopted the static coefficient of friction of the cartilage equal to 0.2.

The FE analysis is implemented assuming that in the initial position, the cartilage is in contact with the cortical region. For each domain of the FE model that is solicited by fractions of the load H, we considered an initial contact pressure equal to the pressure magnitude determined by the boundary load applied.

To analyse the mechanical behaviour of bone tissue for each model, we evaluate the normal and shear components of the stress vector, the maximum shear stress, the principal stress, the normal strain, the shear strain and the principal strain.

We calculated the normal stress as follows (Eq. 7):

$$\sigma_n = n \cdot T \cdot n \tag{7}$$

where n is the normal unit vector to the plane and T is the traction vector.

The shear stress is achieved as the difference between the traction vector and the normal stress vector. Its modulus is obtained as in Eq. 8:

$$\tau = \sqrt{(T_x - \sigma_x)^2 + \left(T_y - \sigma_y\right)^2 + (T_z - \sigma_z)^2} \tag{8}$$

where T_i and σ_i are the components of the traction vector and normal stress vector, respectively.

The maximum shear stress is assessed following the Eq. 9:

$$\tau_{max} = \frac{\sigma_{max} - \sigma_{min}}{2} \tag{9}$$

where σ_{min} and σ_{max} are maximum and minimum principal stresses, respectively.

The normal strain is calculated as in Eq. 10:

$$\varepsilon_n = \varepsilon_{XX} + \varepsilon_{YY} + \varepsilon_{ZZ} \tag{10}$$

where ε_{ii} are the strain tensor components.

The shear strain is obtained from Eq. 11:

$$\varepsilon_n = \varepsilon_{XY} + \varepsilon_{YX} + \varepsilon_{YZ} + \varepsilon_{ZY} + \varepsilon_{XZ} + \varepsilon_{ZX} \tag{11}$$

where ε_{ii} are the strain tensor components.

3 Results and Discussion

In the present study, we developed an FE model of the proximal human femur that mimics the presence of osteophytes and the evolution of OA by altering the load distributions on the femur head. We also considered the influence of the friction phenomenon that occurs at the interface between the cartilage and the cortical tissue. Moreover, with the aim of a more accurate simulation of the physiological conditions, we included in the computational model the muscular forces acting on the great trochanter. The effects of these features on the local mechanical responses of the bone tissue were evaluated during the stance phase of gait (Fig. 2).

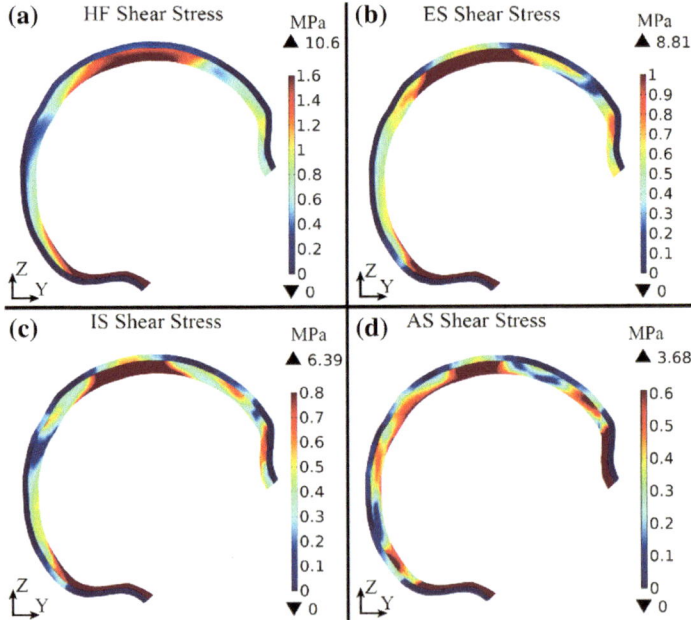

Fig. 2 Shear stress distribution in the coronal plane (YZ plane) at the interface between the cartilage and the cortical tissue of the four models, i.e. healthy case (HF), early stage (ES), intermediate stage (IS) and advanced stage (AS) of the OA femur

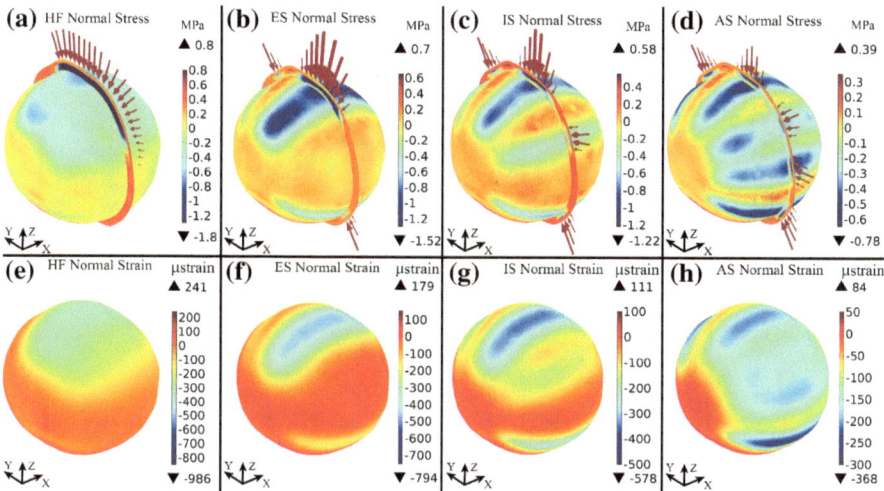

Fig. 3 Normal stress (**a–d**) and normal strain (**e–h**) distribution in the trabecular region of the HF (**a, e**), ES (**b, f**), IS (**c, g**) and AS (**d, h**) models of the femoral head. In Fig. **a–d**, we represent the parallel slice to the YZ plane (coronal plane) that is analysed in Fig. 5. The dark red arrows indicate the load distribution that acts on the femoral head. The arrow length is proportional to the load magnitude

Muscoloskeletal loading influences the stress and the strain within the human femur as bone tissue adapts its microstructure in response to loading [25]. For all healthy and pathological conditions, the outcomes analysed include components and quantities derived from the stress and strain tensors, namely the normal (Fig. 3) and tangential (Fig. 4) components of the stress and the strain, the pattern of principal stress and strain (Fig. 5), the maximum shear stress and the shear strain (Fig. 6). In Figs. 5 and 6, we investigate the quantities of interest in a parallel slice to the anatomical coronal plane, i.e. YZ plane, localized in correspondence of the most solicited region. Thus, we report the distributions of the first principal stress and strain that characterize the coronal slice sited in correspondence to the coordinate system origin, whilst the patterns of the maximum shear stress and the shear strain are critical at a distance $R = 9$ mm on the positive X-axis direction with respect to the coordinate system origin.

Overall, the outcomes are in close agreement with clinical studies [5] and previous FE models [10, 11, 14, 15]. In all pathological scenarios, anomalies of the mechanical behaviour of bone tissue with respect to the HF condition were found in the areas directly under the abnormal load and result from compressive solicitations.

As highlighted in literature [46], the major contribution of the friction condition imposed at the cartilage-cortical bone interface can be assessed by the shear stress (Fig. 2). Elevated values of stress are observed in correspondence of the MCA for all models. With the evolution of the disease, a relocation of stress could be noted, with an increment of the extension of intermediately solicited areas. Generally, the

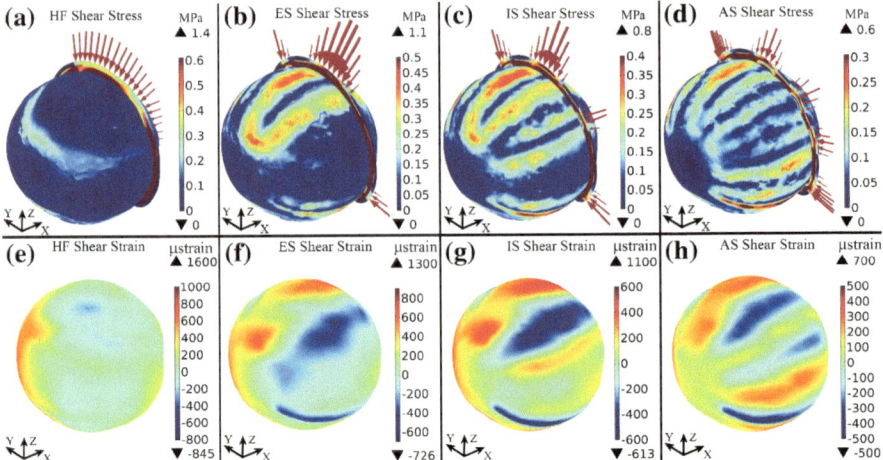

Fig. 4 Shear stress (**a–d**) and shear strain (**e–h**) distribution in the trabecular region of the HF (**a, e**), ES (**b, f**), IS (**c, g**) and AS (**d, h**) models of the femoral head. In Fig. **a–d**, we represent the parallel slice to the YZ plane (coronal plane) that is analysed in Fig. 6. The dark red arrows indicate the load distribution that acts on the femoral head. The arrow length is proportional to the load magnitude

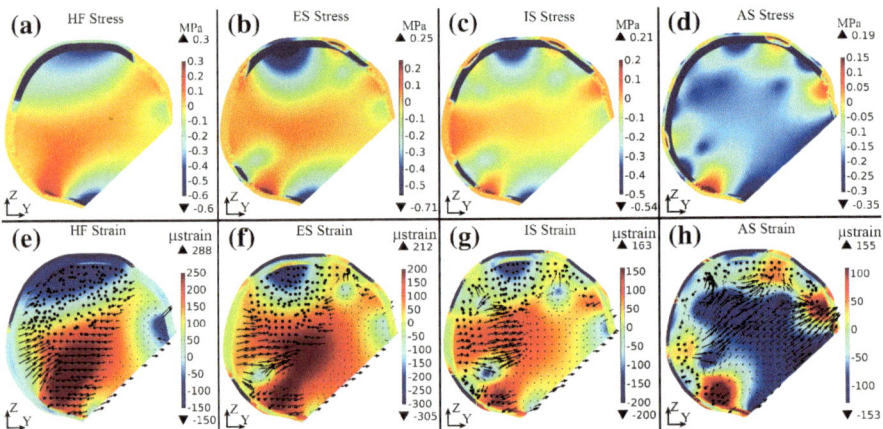

Fig. 5 Principal stress (**a–d**) and principal strain (**e–h**) distribution in the YZ plane (coronal plane) of the HF (**a, e**), ES (**b, f**), IS (**c, g**) and AS (**d, h**) models of the femoral head. In Fig. **e–h,** we represent the vector map of the principal strain. The outcomes are illustrated for the three layers composing the femur head model (cartilage, cortical and trabecular bone)

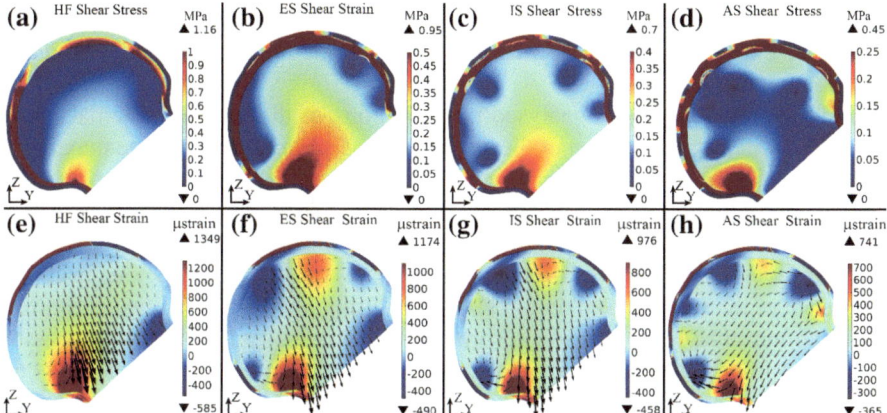

Fig. 6 Maximum shear stress (**a–d**) and shear strain (**e–h**) distribution in the YZ plane (coronal plane) of the HF (**a, e**), ES (**b, f**), IS (**c, g**) and AS (**d, h**) models of the femoral head. In Fig. **e–h**, we represent the vector map of the shear strain. The outcomes are illustrated for the three layers composing the femur head model (cartilage, cortical and trabecular bone)

results lie in the interval 0–10 MPa and are consistent with experimentally studies which confirmed that large shear stresses could be observed at the interface between cartilage and cortical tissue [46].

The distribution and magnitude of stress and strain allow to analyse thoroughly the alteration of the bone tissue. Figure 3 reports the normal stress (a–d) and strain (e–h) that characterize the trabecular region of the femur head. For all loading scenarios considered, in correspondence of the contact points, the tissue is load in compression, while the central region of the femoral head results in tension. The normal stress magnitude lies in the ranges reported in literature [46]. With the evolution of the disease, we observed a decrease of the stress values interval since the load is applied on a greater area composed of the reduced MCA and the areas attributed to the osteophytes presence. Whilst in the HF condition a gradual transition from high to low stress could be observed, in the pathological cases, an alternation of stressed and unstressed regions emerges.

Analogous observations could be performed for the normal strain distribution (Fig. 3 e–h), i.e. tensile strain in the unload areas of the femur head and compressive strain in correspondence of the contact points for each FE model. As expected, with the progression of OA a prominent change in the strain pattern could be noted since the compressive strain area increases its extension.

In terms of trabecular shear stress and strains (Fig. 4), similar patterns with respect to the normal distribution are achieved. The tangential components of the stress and strains are characterized by higher values in comparison to the values obtained in the normal direction. Nonetheless, a slight variation in the shear stress distribution can be observed in the HF model while the OA cases are characterized by a gradual increment of the extension of the stressed area with respect to the low stressed region.

Previously related studies [46] found that shear stress and strain represent quantities of interest that can highlight harmful conditions which contribute actively to the development of OA.

In the coronal plane, we analysed the variations of the second principal stress and strain with the progression of the disease (Fig. 5). For all models, the maximum tensile stress is found at the interface between the cortical and trabecular regions, while overstimulated zones in terms of compressive stress characterizes the MCA and the pathological contact points considered to mimic the presence of osteophytes. Previous models predicted the highest compressive stress to occur in bone in correspondence of the subchondral region [14, 47]. During abnormal loading conditions, the subchondral bone is the main absorber of the impact load. This ability is associated with microcrack formation that could lead to the appearance of altered bone structures, i.e. geodes or eburnations. In comparison with the HF condition, the ES and IS cases show a similar stress pattern with the central trabecular region described by a slight tensile stress. The altered load distribution considered for the AS model leads to the development of a predominant compressive stress pattern in the trabecular zone. The alternated pattern of compressive and tensile stress experienced in the subchondral region for all pathological conditions is comparable with the previous 2D studies [10, 11].

The analysis of the second principal strain leads to similar observations with regard to the pattern features, i.e. a predominant tensile strain in the trabecular regions of HF, ES and IS models, while a compressive strain characterizes the AS trabecular region. Nonetheless, the abnormal load distribution applied to the pathological models implies the development of isolated regions of compressive strain for all models. These zones are also identified by changes in the direction of the strain vector that could imply the presence of anomalous bone structures. Furthermore, the principal strain magnitude determined in this study is comparable to the strain values and distribution achieved by Van Rietbergen et al. [25].

The maximum shear stress (Fig. 6a–d) achieves elevated values in the cortical region, while in the trabecular bone, the highest stress is denoted in the femoral neck region. In the trabecular zone, regions characterized by lower values of shear stress increase their extension with the evolution of OA. According to the previous investigations [10], a high turnover of bone tissue is identified in stimulated zones, while understressed regions represent a favourable site for geodes development [6,13]. Subchondral sclerosis is commonly reported in OA patients and it is widely considered a main feature of advanced OA [12, 13].

In the subchondral trabecular bone, OA is associated with an abnormal low mineralization pattern [2]. The presence of osteophytes leads to a critical load distribution which induces an altered remodelling response. This is well corroborated also by the shear strain (Fig. 6e–h) in the subchondral region. Generally, with the evolution of OA, large variations of the strain vectors direction and magnitude are observed for this zone. Furthermore, the regions characterized by this mechanical behaviour coincide with the zones where clinical observations highlight the development of bone geodes [5].

The previous investigations of Marinozzi et al. [10, 11] and Turmezei et al. [6] support the assumption that osteophytes have a significative impact on the architecture of bone tissue. The abnormal load distribution due to the presence of bone outgrowths leads to the development of regions with alternated low and high values of strain that coincides with the geodes localization identified by Jeffery [5]. The agreement with clinical observations [5, 12] regarding the pattern and the location of bone sclerosis validates the load distribution adopted in the OA model of the femur head.

While these results are promising, the model has some limitations. We considered the thickness of cartilage to be constant in the healthy and OA situations. This assumption is supported by the recent investigation that highlights the existence of bone alterations in the absence of cartilage damage [19]. However, predictions considering different cartilage thickness should be further compared and its importance could be emphasized in such case. We also adopted the hypothesis that bone tissue exhibits linear, elastic, homogeneous, isotropic behaviour. Although this approach is not physiologic, it is frequently used in biomechanical FEM [25] as a first attempt in investigating complex mechanically induced processes in the bone.

4 Conclusions

The 3D FE analysis leads to a model of the mechanical behaviour of the trabecular bone that fits the actual biomechanics measurements which predict regions of high contact pressures at the superior and superolateral femoral head [48]. The evolution of the 3D model confirms the results achieved in the 2D analysis [10, 11] though the present model allows to more accurately assess the strain and stress patterns within the hip joint.

We suggest that the altered load distributions due to the osteophytes can influence the appearance of subchondral bone anomalies. The nature of this mechanism is, however, yet to be clarified and further research is required to confirm this assumption. For instance, a future enhancement of the model should consider non-linear FE simulations. Nevertheless, in the context of the specified limitations, the FE approach presented in this work could provide new insights by identifying potentially harmful defects that could promote progressive trabecular loss. A more sophisticated simulation of muscle attachments, material properties and geometry of osteophytes could enforce the OA knowledge and could provide support for predictive studies of the properties of bone tissue [49] or bone substitutes, especially if combined with biosensors based on ZnO nanomaterials [50, 51].

References

1. Woolf AD, Pfleger B (2003) Burden of major musculoskeletal conditions. Bull World Health Organ 81(9):646–656
2. Li G, Yin J, Gao J, Cheng TS, Pavlos NJ, Zhang C, Zheng MH (2013) Subchondral bone in osteoarthritis: insight into risk factors and microstructural changes. Arthritis Res Ther 15:223
3. Cox LGE, Van Rietbergen B, Van Donkelaar CC, Ito K (2011) Bone structural changes in osteoarthritis as a result of mechanoregulated bone adaptation: a modeling approach. Osteoarthr Cartil 19(6):676–682
4. Goldrin SR (2012) Alterations in periarticular bone and cross talk between subchondral bone and articular cartilage in osteoarthritis. Ther Adv Musculoskelet Dis 4(4):249–258
5. Jeffery AK (1975) Osteophytes and the osteoarthritic femoral head. J Bone Jt Surg Br 57(3):314–324
6. Turmezei TD, Poole KES (2011) Computed tomography of subchondral bone and osteophytes in hip osteoarthritis: the shape of things to come?. Front Endocrinol 2, Article 97
7. Turmezei TD, Lomas DJ, Hopper MA, Poole KES (2014) Severity mapping of the proximal femur: a new method for assessing hip osteoarthritis with computed tomography. Osteoarthr Cartil 22(10):1488–1498
8. Olstza MJ, Cheng X, Jee SS, Kumar R, Kim YY, Kaufman MJ, Douglas EP, Gower LB (2007) Bone structure and formation: a new perspective. Mat Sci Eng R 58:77–116
9. Davidson ENB, Vitters EL, Van Beuningen HM, Van De Loo FAJ, Van Den Berg WB, Van Der Kraan PM (2007) Resemblance of osteophytes in experimental osteoarthritis to transforming growth factor β-induced osteophytes: limited role of bone morphogenetic protein in early osteoarthritic osteophyte formation. Arthritis Rheum 56(12):4065–4073
10. Marinozzi F, Bini F, De Paolis A, De Luca R, Marinozzi A (2015) Effects of hip osteoarthritis on mechanical stimulation of trabecular bone: a finite element study. J Med Biol Eng 35(4):535–544
11. Marinozzi F, Bini F, De Paolis A, Zuppante F, Bedini R, Marinozzi A (2015) A finite element analysis of altered load distribution within femoral head in osteoarthritis. Comput Methods Biomech Biomed Eng Imaging Vis 3(2):84–90
12. Landells JW (1953) The bone cysts of osteoarthritis. J Bone Joint Surg 35B(4):643–649
13. Resnick D, Niwayama G, Coutts RD (1977) Subchondral cysts (geodes) in arthritic disorders: pathologic and radiographic appearance of the hip joint. Am J Roentgenol 128(5):799–806
14. Bini F, Pica A, Marinozzi A, Marinozzi F (2018) Prediction of osteophytes relevance in human osteoarthritic femur head from load pattern rearrangement simulations: an integrated fem study. In: Proceedings of the 15th international symposium on computer methods in biomechanics and biomedical engineering and 3rd conference on imaging and visualization CMBBE
15. Duda GN, Heller M, Albinger J, Schulz O, Schneider E, Claes L (1998) Influence of muscle forces on femoral strain distribution. J Biomech 31(9):841–846
16. Stolk J, Verdonschot N, Huiskes R (2001) Hip-joint and abductor-muscle forces adequately represent in vivo loading of a cemented total hip reconstruction. J Biomech 34(7):917–926
17. Marinozzi F, Marinozzi A, Bini F, Zuppante F, Pecci R, Bedini R (2012) Variability of morphometric parameters of human trabecular tissue from coxo-arthritis and osteoporotic samples. Ann I Super Sanità 48(1):19–25
18. Marinozzi F, Bini F, Marinozzi A, Zuppante F, De Paolis A, Pecci R, Bedini R (2013) Technique for bone volume measurement from human femur head samples by classification of micro-CT image histograms. Ann I Super Sanità 49(3):300–305
19. van der Kraan PM, van den Berg WB (2007) Osteophytes: relevance and biology. Osteoarthr Cartil 15(3):237–244
20. Bini F, Marinozzi A, Marinozzi F, Patanè F (2002) Microtensile measurements of single trabeculae stiffness in human femur. J Biomech 35(11):1515–1519
21. Marinozzi F, Bini F, Marinozzi A (2011) Evidence of entropic elasticity of human bone trabeculae at low strains. J Biomech 44(5):988–991

22. Bini G, Bini F, Bedini R, Marinozzi A, Marinozzi F (2017) A topological look at human trabecular bone tissue. Math Biosci 288:159–165
23. Pustoc'h A, Cheze L (2009) Normal and osteoarthritic hip joint mechanical behaviour: a comparison study. Med Biol Eng Comput 47(4):375–383
24. Cowin SC (1999) Bone poroelasticity. J Biomech 32(3):217–238
25. Van Rietbergen B, Huiskes R, Eckstein F, Ruegsegger P (2003) Trabecular bone tissue strains in the healthy and osteoporotic human femur. J Bone Min Res 18:1781–1788
26. Verhulp E, van Rietbergen B, Huiskes R (2008) Load distribution in the healthy and osteoporotic human proximal femur during a fall to the side. Bone 42(1):30–35
27. Marinozzi F, De Paolis A, De Luca R, Bini F, Bedini R, Marinozzi A (2012) Stress and strain patterns in normal and osteoarthritic femur using finite element analysis. In: Proceedings of compimage-computational modelling of objects represented in images iii: fundamentals, methods and applications, pp 247–250
28. Richard F, Villars M, Thibaud S (2013) Viscoelastic modeling and quantitative experimental characterization of normal and osteoarthritic human articular cartilage using indentation. J Mech Behav Biomed Mater 24:41–52
29. Athanasiou KA, Rosenwasser MP, Buckwalter JA, Malinin TI, Mow VC (1991) Interspecies comparisons of insitu intrinsic mechanical-properties of distal femoral cartilage. J Orthop Res 9(3):330–340
30. Miyasaka D, Ito T, Imai N, Suda K, Minato I, Dohmae Y, Endo N (2014) Three-dimensional assessment of femoral head coverage in normal and dysplastic hips: a novel method. Acta Med Okayama 68(5):277–284
31. Miller Z, Fuchs MB, Arcan M (2002) Trabecular bone adaptation with an orthotropic material model. J Biomech 35(2):247–256
32. Genda E, Konishi N, Hasegawa Y, Miura T (1995) A computer simulation study of normal and abnormal hip joint contact pressure. Arch Orthop Trauma Surg 114(4):202–206
33. Bergmann G, Deuretzbacher G, Heller M, Graichen F, Rohlmann A, Strauss J, Duda GN (2001) Hip contact forces and gait patterns from routine activities. J Biomech 34:859–871
34. Brinckmann P, Frobin W, Hierholzer E (1981) Stress on the articular surface of the hip joint in healthy adults and persons with idiopathic osteoarthrosis of the hip joint. J Biomech 14(3):149–156
35. Bachtar F, Chen X, Hisad T (2006) Finite element contact analysis of the hip joint. Med Biol Eng Comput 44(8):643–651
36. Johnson KL (1985) Contact mechanics. Cambridge University Press, Cambridge
37. Greenwald AS, O'Connor JJ (1971) The transmission of load through the human hip joint. J Biomech 4(6):507–528
38. Naghieh GR, Jin ZM, Rahnejat H (1998) Contact characteristics of viscoelastic bonded layers. Appl Math Model 22(8):569–581
39. Fang X, Zhang C, Chen X, Wang Y, Tan Y (2015) A new universal approximate model for conformal contact and non-conformal contact of spherical surfaces. Acta Mech 226(6):1657–1672
40. Pompe B, Daniel M, Sochor M, Vengust R, Kralj-Iglič V, Iglič A (2003) Gradient of contact stress in normal and dysplastic human hips. Med Eng Phys 25(5):379–385
41. Heller MO, Bergmann G, Kassi JP, Claes L, Haas NP, Duda GN (2005) Determination of muscle loading at the hip joint for use in pre-clinical testing. J Biomech 38(5):1155–1163
42. Venäläinen MS, Mononen ME, Salo J, Räsänen LP, Jurvelin JS, Toyräs J, Viren T, Korhonen RK (2016) Quantitative evaluation of the mechanical risks caused by focal cartilage defects in the knee. Sci Rep 6:1–11
43. Banijamali SMA, Oftadeh R, Nazarian A, Goebel R, Vaziri A, Nayeb-Hashemi H (2015) Effects of different loading patterns on the trabecular bone morphology of the proximal femur using adaptive bone remodeling. J Biomech Eng 137(1):011011-1-8
44. Brand RA, Pedersen DR, Friederich JA (1986) The sensitivity of muscle force predictions to changes in physiologic cross-sectional area. J Biomech 19(8):589–596
45. Lizhang J, Fisher J, Jin Z, Burton A, Williams S (2011) The effect of contact stress on cartilage friction, deformation and wear. Proc Inst Mech Eng Part H J Eng Med 225(5):461–475

46. Ateshian GA, Henak CR, Weiss JA (2015) Toward patient-specific articular contact mechanics. J Biomech 48(5):779–786
47. Malekipour F, Oetomo D, Lee PVS (2017) Subchondral bone microarchitecture and failure mechanism under compression: a finite element study. J Biomech 55:85–91
48. Turmezei TD, Treece GM, Gee AH, Fotiadou AF, Poole KES (2016) Quantitative 3D analysis of bone in hip osteoarthritis using clinical computed tomography. Eur Radiol 26(7):2047–2054
49. Bini F, Pica A, Marinozzi A, Marinozzi F (2017) 3D diffusion model within the collagen apatite porosity: an insight to the nanostructure of human trabecular bone. PLoS One 12(12)
50. Araneo R, Rinaldi A, Notargiacomo A, Bini F, Pea M, Celozzi S, Marinozzi F, Lovat G () Design concepts, fabrication and advanced characterization methods of innovative piezoelectric sensors based on ZnO nanowires. Sensors 14(12):23539–23562
51. Araneo R, Rinaldi A, Notargiacomo A, Bini F, Marinozzi F, Pea M, Lovat G, Celozzi S (2014) Effect of the scaling of the mechanical properties on the performances of ZnO piezo-semiconductive nanowires. In: AIP conference proceedings, vol 1603, pp 14–22

Numerical Simulation of the Deployment Process of a New Stent Produced by Ultrasonic-Microcasting: The Role of the Balloon's Constitutive Modeling

I. V. Gomes, H. Puga and J. L. Alves

Abstract The application of the Finite Element Method (FEM) allows to predict the behavior of a stent during the deployment process and when in service, being a powerful tool to use in its design and development. As the promoter of the stent expansion, the balloon plays a very important role, offering a strong influence on its performance, mainly during the deployment process. This element is usually built in a rubber-like material such as polyurethane, being modeled as linear elastic or hyperelastic with a Mooney–Rivlin description. This work aims, through FEM analysis, the study of the influence of both adopted material formulation—linear elastic or hyperelastic—as well as the respective material constants and properties for the balloon modeling on the performance of a biocompatible magnesium stent regarding a set of metrics. Furthermore, a comparison is established between those results and the obtained ones in the scenario of application of pressure directly in the inner surface of the stent, neglecting the balloon. The obtained results suggest that material formulation has a direct influence on the stent deployment process. Concerning to hyperelastic models, two different combinations of parameter values were tested, showing a similar behavior regarding the considered metrics, while the linear elastic model presents comparable values for the expansion pressure and recoil, but different in terms of dogboning and foreshortening. The scenario of neglecting the balloon suggests providing the highest values of dogboning, foreshortening, and recoil, with an expansion pressure inferior to that of hyperelastic models.

I. V. Gomes · H. Puga · J. L. Alves
CMEMS – Center for Microelectromechanical Systems, Guimarães, Portugal

I. V. Gomes (✉)
MIT-Portugal Program, Department of Mechanical Engineering, University of Minho, Campus of Azurém, 4800-058 Guimarães, Portugal
e-mail: inesvarelagomes@gmail.com

© Springer Nature Switzerland AG 2019
J. M. R. S. Tavares and P. R. Fernandes (eds.), *New Developments on Computational Methods and Imaging in Biomechanics and Biomedical Engineering*, Lecture Notes in Computational Vision and Biomechanics 33, https://doi.org/10.1007/978-3-030-23073-9_5

1 Introduction

Coronary heart diseases such as atherosclerosis are, nowadays, one of the major causes of death in the world [1]. One of the possible treatments for this condition is the deployment of a stent, a tiny wire mesh tube-like structure, which is radially expanded through the inflation of a balloon placed within it, reopening the vessel and acting as a scaffold [2, 3].

The Finite Element Method (FEM) is a powerful tool used in the study, design, and development of these devices once it allows to predict their behavior in a more expeditious way with lower costs when compared to the experimental techniques [4]. Thus, to guarantee that the obtained results are trustworthy, it is crucial the correct definition of the system, namely in terms of the material constitutive models applied to the involved constituents, including the stent and the balloon.

As the promoter of the stent expansion, the balloon plays an important role, having a strong influence on its performance, mainly during the deployment process. The balloon element is made of rubber-like materials such as polyurethane [4–7] or nylon [8] and its constitutive behavior is modeled through different formulations, being the linear elastic and the hyperelastic with Mooney–Rivlin description the most used. Such definition is suggested to have an impact on the results obtained by the analysis by FEM and therefore, on their applicability to the reality. Wang et al. [9], Gervaso et al. [10] and Pant et al. [11] have adopted the linear elastic model for cylindrical-shaped balloons, being this formulation more common when a folded balloon geometry is used as presented by Schiavone et al. [5, 12, 13] and De Beule et al. [6]. Although the rubber-like materials are characterized by its incompressibility or near-incompressibility and highly nonlinear behavior at large strains, in the region of small strains, which is characterized by a linear behavior, a Young's modulus may be assigned [14]. Hyperelastic models with Mooney–Rivlin description were preferred for cylindrical balloons modeling by Chua et al. [15], Schiavone et al. [5], Eshghi et al. [7] and Beigzadeha et al. [16] with different values for material parameters. The material constants are derived from experimental data of mechanical tests, where different tests may lead to different values. Consequently, the mechanical tests to perform to provide information for the material constants calculation must represent a stress state as close as possible to the one that the studied element expectedly undergoes to describe the material behavior in a reliable way. The aforementioned relation between geometry and material constitutive modeling and its impact on the trustworthiness of the results obtained by Finite Element Analysis (FEA) highlights these two factors as of major importance and therefore, as deserving a study focused on it.

Hence, the present work aims the study of the influence of the adopted material formulation—linear elastic or hyperelastic—and respective material constants and properties for the balloon modeling on the performance of a biocompatible magnesium alloy stent, using FEM method. Furthermore, a comparison is established between those results and the obtained ones in the scenario of application of pressure directly in the inner surface of the stent to mimic the presence of the balloon.

2 Description of the Methodology

In the present work, the influence of balloon material constitutive modeling is assessed using two different material formulations—linear elastic and hyperelastic with Mooney–Rivlin description, being used two different sets of material parameters, designated as Hyperelastic 1 (HE1) and Hyperelastic 2 (HE2).

A simpler simulation is performed neglecting the presence of the balloon. In this case, the stent expansion is promoted by the application of a linear pressure directly in the inner surface of the stent until the target diameter is reached, after what the stent slightly recoils by the absence of applied load, what mimics the deflation of the balloon. When considering the inflation of the balloon within the stent, a linear pressure is applied on its inner surface, leading to its expansion and consequently, to the radial deformation of the stent. An augmented Lagrange formulation is adopted for the contact between the stent and the balloon, which is considered frictionless. This method allows to produce less penetration and better accuracy than the pure penalty method presenting, however, higher computational cost.

2.1 Geometry

The NG stent geometry [17] results of a combination of straight lines and arcs, which are expected to contribute to both reduction of the expansion pressure and foreshortening phenomenon. The linkage between the rings of the structure is made using curved "bridge" elements, whose deformation during the expansion process allows the compensation of the length reduction due to the radial expansion. The nonexpanded inner diameter of the stent and its length is 5 and 24 mm, respectively, and its thickness is equal to 0.1 mm.

This geometry presents periodicity in the circumferential and longitudinal directions in the cylindrical coordinate system as it is formed by the assembling of multiple identical unit cells. Along with the geometrical periodicity, the isotropic behavior of the material allows the use of only one-tenth (1/5 in the circumferential direction and 1/2 in the longitudinal direction) of the model, reducing the required computation time thanks to a smaller number of Degrees of Freedom (DOFs) while the accuracy of the results is guaranteed. For this purpose, symmetry conditions are applied to all degrees of freedom of the stent belonging to planar faces α, β, and γ, as shown in Fig. 1a.

A cylindrical balloon with open ends is used to promote the radial expansion of the stent, presenting an inner diameter equal to 4.7 mm, thickness of 0.15 mm, and a full length of 26 mm. Once again, due to the geometrical symmetry and isotropic behavior of the material, only one-tenth of the balloon model is used, being applied symmetry conditions on all its degrees of freedom that belong to the planar faces α, β, and γ, as presented in Fig. 1a, being the ends fully constrained to represent the

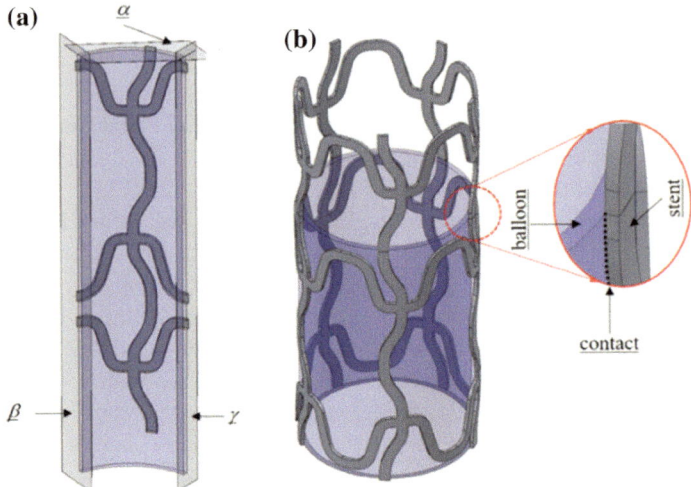

Fig. 1 Balloon and stent set **a** identification of the planes of symmetry; **b** assembly of balloon and stent with identification of the contact area

bond to the catheter. The assembly of both elements (stent and balloon) is presented in Fig. 1b.

2.2 Material Constitutive Modeling

The material selected for the stent is a magnesium alloy whose mechanical behavior, due to the manufacturing process, is assumed to be isotropic and elastoplastic with Young's modulus $E = 43$ GPa and Poisson's ratio $v = 0.30$, being the plastic behavior modeled by the von Mises yield criterion. The nonlinear isotropic stress–strain hardening curve is modeled by a Voce-type law defined by Eq. (1).

$$\sigma = \sigma_0 + (\sigma_{sat} - \sigma_0)(1 - \exp(-C_y \times \bar{\varepsilon}^p)) \tag{1}$$

where σ_0 is the initial yield stress, equal to 174.8 MPa, σ_{sat} is the saturation flow stress, equal to 315.6 MPa, C_y is the unitless hardening rate, equal to 16.3 and $\bar{\varepsilon}^p$ is the equivalent plastic strain, work conjugate of the von Mises equivalent stress.

The constitutive modeling of balloon hyperelastic material, characterized by low elastic modulus and high bulk modulus, is derived from the strain energy function (W), which represents the energy stored in the material per unit of reference volume, based on three strain invariants of the Cauchy–Green deformation tensor ($\bar{I}_1$, $\bar{I}_2$ and $\bar{I}_3$). Once hyperelastic materials are considered incompressible, $\bar{I}_3$ is equal to 1 and W becomes a function dependent only on $\bar{I}_1$ and $\bar{I}_2$.

Table 1 Material constants for hyperplastic material models

Material model	C_{10} [MPa]	C_{01} [MPa]	k[MPa]
HE1	1.03 [5]	3.69 [5]	10^4
HE2	−0.89	5.39	

Table 2 Elastic properties of linear elastic material models

Material model	E [MPa]	v [−]	ρ [kg/m3]
LE	10 [9]	0.49 [9]	1100 [6]

Mooney–Rivlin models are one of the possible formulations for the description of the behavior of hyperelastic materials, whose generic expression is given by Eq. (2).

$$W = \sum_i \sum_j C_{ij}(\bar{I}_1 - 3)^i (\bar{I}_2 - 3)^j + \frac{1}{2}k(J_{el} - 1)^2 \qquad (2)$$

where $\bar{I}_1$ and $\bar{I}_2$ are the first and second invariant of the left isochoric Cauchy–Green deformation tensor, J_{el} is the elastic Jacobian, k is the bulk modulus, and C_{ij} are model parameters, resulting from material experimental data from mechanical tests. In this study, and according to Eq. (3), a two-parameter Mooney–Rivlin model to describe the behavior of polyurethane rubber is used, being the values of the material constants presented in Table 1.

$$W = C_{10}(\bar{I}_1 - 3) + C_{01}(\bar{I}_2 - 3) + \frac{1}{2}k(J_{el} - 1)^2 \qquad (3)$$

The values of the material constants of HE2 model were obtained through curve fitting of experimental data from a uniaxial tensile test performed on a polyurethane sample.

Regarding the linear elastic behavior modeling of balloon material, the material properties used in the study are presented in Table 2.

3 Results and Discussion

Figure 2 presents the required expansion pressure as a function of the expanded radius of the stent.

The use of LE model shows results close to those obtained through the adoption of HE1 and HE2 models, being the difference equal to approximately 5.13% (0.39 MPa vs. 0.37 MPa) for the first and 10.26% (0.39 MPa vs. 0.35 MPa) for the second. According to the aforesaid, it is suggested that the adoption of a linear elastic approach to the constitutive modeling of a hyperelastic material for a cylindrical balloon may be a valid option as a similar behavior was found between the LE, HE1, and HE2 models, as demonstrated by Z. Guo et al. [14].

Fig. 2 Required expansion pressure as a function of the expanded radius of the stent

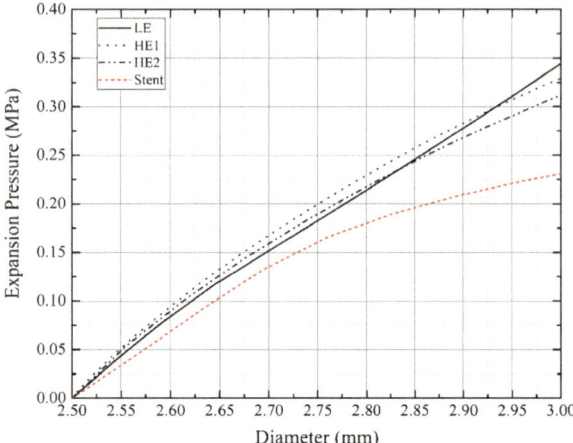

Fig. 3 Evolution of dogboning parameter as a function of the expansion pressure

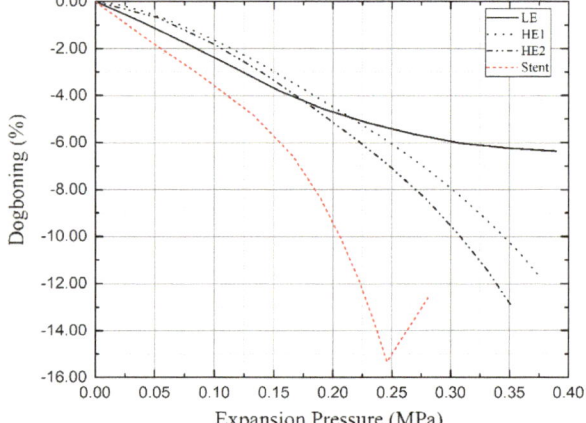

The simulation of the stent itself leads to values of required expansion pressure considerably inferior to those of HE1, HE2, and LE models, about 28.00%, suggesting that such simplification may not provide adequate results despite the minor computational cost.

Figure 3 presents the performance of dogboning parameter, whose evolution is a function of the expansion pressure of the balloon.

All the cases studied present negative values of dogboning, which means that the central region of the stent experiences greater expansion than its ends. Moreover, the models present an increase of dogboning parameter as the expansion pressure rises excepting the direct pressurization of the stent, being the hyperelastic models (HE1 and HE2) the ones that present higher values with post-dilatation absolute values of 11.60 and 13.02%, respectively. When the presence of the balloon is neglected, the dogboning effect presents a different behavior comparing to that of other scenarios,

Fig. 4 Evolution of
foreshortening parameter as
a function of the expansion
pressure

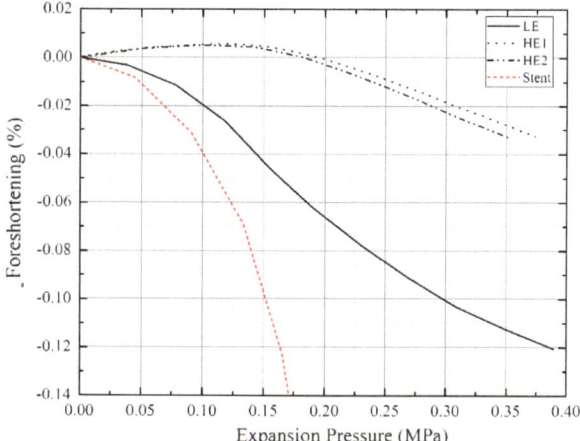

as its maximum value is not reached at full expansion. Once the pressure application
is ceased as the target diameter is reached in each point, the recoil phenomenon does
not occur simultaneously for all the structure, resulting in that the central section of
the stent experiences it first than its ends. The referred situation leads to the presented
inversion of the dogboning tendency.

Conversely to what is verified for the relation between the applied expansion
pressure and the stent diameter, the dogboning effect suggests being more sensitive
to the adoption of a linear model instead of a hyperelastic one once more significant
differences are found in the results.

The results obtained in foreshortening evaluation are presented in Fig. 4. The
results show that the adoption of hyperelastic formulations for modeling the constitutive behavior of balloon leads to an elongation (positive values of foreshortening)
of the stent in the initial phase of the expansion changing to its shortening (negative
values of foreshortening), which are in agreement with the remaining models. The
linear elastic model LE presents significantly higher values of foreshortening when
compared to those of hyperelastic (HE1 and HE2) models (-0.12% vs. -0.033%),
that are approximately equal. This evidence highlights once again the influence of the
adopted material formulation in the behavior of the system. As verified for dogboning, when a pressure is applied directly in the inner surface of the stent, the maximum
absolute value of foreshortening (-0.53%) does not occur at its maximum diameter,
which may have origin on the aforementioned mechanism.

The different behavior presented by the Linear Elastic (LE) and hyperelastic (HE1
and HE2) models can be promoted by the contact formulation between the stent
and the balloon, as there are related variables strongly dependent on them. Indeed,
the characteristic stiffness of the contact is dependent on the equivalent Young's
modulus, whose value is equal to the Young's modulus itself in the case of linear
elastic models while it depends on the bulk modulus and on the Mooney–Rivlin
constants for hyperelastic formulations, giving origin to different values.

Table 3 Recoil evaluation

	Longitudinal recoil (%)	Distal recoil (%)	Central recoil (%)
LE	−0.08	2.98	6.22
HE1	0.002	2.95	7.12
HE2	0.003	2.86	7.33
Stent	−0.36	5.58	9.27

After the complete expansion of the stent promoted by the balloon inflation, it is deflated in order to be extracted along with the further elements used in the deployment process. At this stage, the pressure exerted by the balloon in the inner surface of the stent ceases and the elastic portion of its deformation is recovered producing a phenomenon called recoil, which can be longitudinal, central and distal. The obtained values for these metrics are presented in Table 3.

Concerning to the longitudinal recoil, which corresponds to the variation of the stent length as the balloon is deflated, all cases studied with exception of the hyperelastic models show negative values and therefore, a shrinkage of the device. The values of longitudinal recoil presented by HE1 and HE2 models are significantly inferior to those of the remaining scenarios and close to zero, meaning that there is no relevant alteration of the length of the stent due to its elastic recovery. The obtained value for the linear elastic model is lower than that of the simulation of the stent without balloon, which leads to the highest absolute value (−0.36%).

The distal recoil corresponds to the difference between the stent diameter in its ends in the fully expanded configuration and after the balloon deflation. Regarding this parameter, it is noticeable that the LE, HE1, and HE2 models present comparable values, whilst the highest one is the obtained through the neglection of the balloon (5.58%). Indeed, the LE value is only about 1.02% superior to that of HE1, while this difference is about 46.59% compared to the use of the stent itself.

The difference between the stent diameter in its mid-region in the aforementioned conditions is called central recoil. The obtained results show that this phenomenon is more relevant than the distal recoil, once higher values are achieved, existing a greater reduction of the stent expanded diameter in the central part. Regarding this parameter, the value presented by the stent itself (9.27%) is the highest one, being significantly above those presented by the remaining models.

The occurrence of such phenomena is a situation with negative impact on the performance of the stent once it may demand an overexpansion of the device in order to take into consideration its diameter loss after the deflation of the balloon, which can compromise the integrity of the blood vessel and therefore, the success of the procedure.

4 Conclusions

In the present work, a systematic study of the influence of the adopted material formulation for the expansion balloon modeling was performed. The main conclusions to be drawn from this study are:

1. the choice of the material constitutive model of the expansion balloon suggests having a significant influence on the stent deployment process simulation.
2. the adoption of different material formulations leads to different results of the considered performance metrics, possibly driving to significant discrepancies between the results obtained through numerical analysis and those then presented by the device in real context. Such situation may have as consequence the inadequacy of the numerical model to correctly describe the behavior of the system, not providing results close to reality and therefore, invalidating the stent's designs proposed by it.
3. the adoption of a linear elastic model may be suitable for the description of the relation between expansion pressure and expanded radius for a cylindrical balloon built in a rubber-like material if a low Young's modulus is assigned. Concerning to the evaluation of the remaining performance metrics, this option does not suggest being a so-appropriate approach, as more significant differences relatively to hyperelastic models are noticed.
4. the direct application of pressure in the inner surface of the stent produces values inferior to linear elastic and hyperelastic models of required expansion pressure, being the remaining metrics significantly higher.

Acknowledgements This work was supported by FEDER funds through the COMPETE program with the reference project PTDC/SEM-TEC/3827/2014 and PTDC/EMS-TEC/0702/2014. Additionally, this work was supported by FCT with the reference project UID/EEA/04436/2013 and by FEDER funds through the COMPETE 2020 with the reference project POCI-01-0145-FEDER-006941.

References

1. Roy T, Chanda A (2014) Computational modeling and analysis of latest commercially available coronary stents during deployment. Procedia Mater Sci 5:2310–2319
2. Li N, Gu Y (2005) Parametric design analysis and shape optimization of coronary arteries stent structure. In: Proceeding 6th World Congress of Structural and Multidisciplinary Optimization, vol 30
3. Azaouzi M, Makradi A, Belouettar S (2013) Numerical investigations of the structural behavior of a balloon expandable stent design using finite element method. Comput Mater Sci 72:54–61
4. Imani M, Goudarzi AM, Ganji DD, Aghili AL (2013) The comprehensive finite element model for stenting: the influence of stent design on the outcome after coronary stent placement. J Theor Appl Mech 51(3):639–648
5. Schiavone A, Zhao LG (2015) A study of balloon type, system constraint and artery constitutive model used in finite element simulation of stent deployment. Mech Adv Mater Mod Process 1(1)

6. De Beule M, Mortier P, Carlier SG, Verhegghe B, Van Impe R, Verdonck P (2008) Realistic finite element-based stent design: The impact of balloon folding. J Biomech 41(2):383–389

7. Eshghi N, Hojjati MH, Imani M, Goudarzi AM (2011) Finite element analysis of mechanical behaviors of coronary stent. Procedia Eng 10:3056–3061

8. Gay M, Zhang L, Liu WK (2006) Stent modeling using immersed finite element method. Comput Methods Appl Mech Eng 195(33–36):4358–4370

9. Wang W-Q, Liang D-K, Yang D-Z, Qi M (2006) Analysis of the transient expansion behavior and design optimization of coronary stents by finite element method. J Biomech 39(1):21–32

10. Gervaso F, Capelli C, Petrini L, Lattanzio S, Di Virgilio L, Migliavacca F (2008) On the effects of different strategies in modeling balloon-expandable stenting by means of finite element method. J Biomech 41(6):1206–1212

11. Pant S, Bressloff NW, Limbert G (2012) Geometry parameterization and multidisciplinary constrained optimization of coronary stents. Biomech Model Mechanobiol 11(1–2):61–82

12. Schiavone A, Zhao LG (2016) A computational study of stent performance by considering vessel anisotropy and residual stresses. Mater Sci Eng C 62:307–316

13. Schiavone A, Abunassar C, Hossainy S, Zhao LG (2016) Computational analysis of mechanical stress–strain interaction of a bioresorbable scaffold with blood vessel. J Biomech 49(13):2677–2683

14. Guo Z, Sluys LJ (2008) Constitutive modeling of hyperelastic rubber-like materials. HERON 53:3

15. Chua SND, Mac Donald BJ, Hashmi MSJ (2003) Finite element simulation of stent and balloon interaction. J Mater Process Technol 143–144:591–597

16. Beigzadeh B, Mirmohammadi SA, Ayatollahi MR (2017) A numerical study on the effect of geometrical parameters and loading profile on the expansion of stent. Biomed Mater Eng 28(5):463–476

17. Gomes IV, Puga H, Alves JL (2017) Shape and functional optimization of biodegradable magnesium stents for manufacturing by ultrasonic-microcasting technique. Int J Interact Des Manuf IJIDeM

18. Li H et al (2017) Design optimization of stent and its dilatation balloon using kriging surrogate model. Biomed. Eng OnLine 16(1)

19. Zahedmanesh H, John Kelly D, Lally C (2010) Simulation of a balloon expandable stent in a realistic coronary artery—Determination of the optimum modeling strategy. J Biomech 43(11):2126–2132

New Techniques for Combined FEM-Multibody Anatomical Simulation

John E. Lloyd, Antonio Sánchez, Erik Widing, Ian Stavness, Sidney Fels, Siamak Niroomandi, Antoine Perrier, Yohan Payan and Pascal Perrier

Abstract This article describes a number of new techniques useful for the construction of biomechanical and anatomical models, particularly those that employ combined FEM-multibody simulation. They are being introduced to the ArtiSynth mechanical modeling system, and include reduced coordinate modeling, in which an FEM model is made more computationally efficient by reducing it to a low degree-of-freedom subspace; new methods for connecting points and coordinate frames directly to deformable bodies; and the ability to create skin and embedded meshes that are connected to underlying FEM models and other dynamic components. All these techniques are based on the principle of virtual work, and we illustrate their application with a number of examples, including a reduced FEM tongue model, subject-specific skeletal registration, skinning applied to modeling the human airway, and a detailed model of the human masseter.

1 Introduction

Effective simulation of human anatomical structure and function can benefit from combining low-fidelity models with fast computation times and high-fidelity models that emulate detailed tissue dynamics but have slower computation times. Multibody methods are typically used for the former, modeling structures such as bones, joints and point-to-point muscles, while finite element methods (FEM) are typically used for the latter, modeling deformable tissues and capturing internal stress/strain dynamics.

J. E. Lloyd (✉) · A. Sánchez · S. Fels
Electrical and Computer Engineering, University of British Columbia, Vancouver, Canada
e-mail: lloyd@ece.ubc.ca

I. Stavness · E. Widing
Computer Science, University of Saskatchewan, Saskatoon, Canada

S. Niroomandi · A. Perrier · Y. Payan
TIMC, Université Joseph Fourier, Grenoble, France

P. Perrier
Institut de la Communication Parlée, INGP & Université Grenoble Alpes, Grenoble, France

© Springer Nature Switzerland AG 2019   75
J. M. R. S. Tavares and P. R. Fernandes (eds.), *New Developments on Computational Methods and Imaging in Biomechanics and Biomedical Engineering*, Lecture Notes in Computational Vision and Biomechanics 33, https://doi.org/10.1007/978-3-030-23073-9_6

Combining the two can enable the creation of models with efficient, and possibly interactive, simulation times while also providing appropriate fidelity in an area of interest.

In this article, we describe new techniques that are being introduced into ArtiSynth [12] (www.artisynth.org), an open- source simulation platform that permits researchers to combine multibody and FEM techniques and hence leverage the advantages of both. These new techniques include reduced coordinate modeling, attaching points and frames to deformable bodies, and skinning and embedded meshes.

2   Reduced Coordinate Modeling

Reduced coordinate modeling is a technique in which a deformable body is modeled using a restricted deformation basis instead of a collection of deformable finite elements [10]. It spans the gap between FEM methods and rigid bodies (which are themselves reduced models condensed to purely rigid motions), and can be effective in speeding up simulation times for models in which the range of typical deformations is constrained (such as tongue motions in speech production).

To create a reduced coordinate model, it is often convenient to begin with a standard FEM model. One can then construct a *basis* $\mathbf{U}$ of nodal deformations (with respect to the nodal rest positions) which spans the set of all possible deformations for the reduced model. Assume the FEM modal has n nodes (each with 3 degrees of freedom), and let $\mathbf{x}$, $\mathbf{x}_0$ and $\mathbf{u}$ denote composite vectors of their positions, rest positions, and displacements, such that $\mathbf{x} = \mathbf{x}_0 + \mathbf{u}$. Then if $\mathbf{q}$ is a vector of the r reduced coordinates, we have

$$\mathbf{x} = \mathbf{x}_0 + \mathbf{U}\mathbf{q}, \tag{1}$$

where $\mathbf{U} \in \mathbb{R}^{3n \times r}$. The basis $\mathbf{U}$ does not have to be constant but often is and will be assumed to be for the remainder of this article. Determining an appropriate basis is one of the principal challenges in constructing a reduced model. Automatic techniques include linear modal analysis [14] (when the deformation is small), along with various ways to extend a modal basis with additional vectors to handle large deformations, such as using modal derivatives [3] or applying additional linear transformations to the basis vectors [22]. In practice, better results are often obtained by creating the basis via a training method in which a non-reduced FEM model is used to recreate the deformations that are required for the modeling application [10].

2.1   Reduced Dynamics

Background material for reduced dynamics modeling can be found in [20]. Here, we provide an overview of reduced dynamics within the ArtiSynth modeling framework. An FEM model advances in time according to the dynamics

$$\mathbf{M}\ddot{\mathbf{x}} + \mathbf{D}\dot{\mathbf{x}} + \mathbf{K}\delta\mathbf{x} = \mathbf{f}_{\text{int}}(\mathbf{x}) + \mathbf{f}_{\text{ext}}, \tag{2}$$

where $\mathbf{M}$ and $\mathbf{D}$ are mass and damping matrices, $\mathbf{K}$ is the local stiffness matrix, $\delta\mathbf{x}$ is the local change in $\mathbf{x}$, and $\mathbf{f}_{\text{int}}$ and $\mathbf{f}_{\text{ext}}$ are the internal and external forces. Note that the matrices in (2) are almost always sparse.

ArtiSynth uses (2), in conjunction with a semi-implicit integrator, to solve for the motion of the FEM model. In order to handle reduced models, it is necessary to find the equivalent reduced dynamics,

$$\tilde{\mathbf{M}}\ddot{\mathbf{q}} + \tilde{\mathbf{D}}\dot{\mathbf{q}} + \tilde{\mathbf{K}}\delta\mathbf{q} = \tilde{\mathbf{f}}_{\text{int}}(\mathbf{q}) + \tilde{\mathbf{f}}_{\text{ext}}, \tag{3}$$

where $\tilde{\mathbf{M}}$, $\tilde{\mathbf{D}}$, and $\tilde{\mathbf{K}}$ are the reduced mass, damping, and stiffness matrices, and $\tilde{\mathbf{f}}_{\text{int}}$ and $\tilde{\mathbf{f}}_{\text{ext}}$ are the reduced internal and external forces. Note that all of the matrices in (3) are dense.

Model reduction implies a linear relationship between the nodal velocities $\dot{\mathbf{x}}$ of the original FEM model and the velocities $\dot{\mathbf{q}}$ of the reduced model:

$$\dot{\mathbf{x}} = \mathbf{U}\dot{\mathbf{q}}.$$

Note that even if $\mathbf{U}$ were not constant, this would still be true locally. Then from the principle of virtual work, we know that the work done in nodal coordinates, $\mathbf{f}^T\dot{\mathbf{x}}$, must equal the work done in reduced coordinates, $\tilde{\mathbf{f}}^T\dot{\mathbf{q}}$, and therefore

$$\tilde{\mathbf{f}} = \mathbf{U}^T\mathbf{f}.$$

This allows us to determine the reduced quantities in (3):

$$\tilde{\mathbf{M}} = \mathbf{U}^T\mathbf{M}\mathbf{U}, \quad \tilde{\mathbf{D}} = \mathbf{U}^T\mathbf{D}\mathbf{U}, \quad \tilde{\mathbf{K}} = \mathbf{U}^T\mathbf{K}\mathbf{U}, \quad \tilde{\mathbf{f}}_{\text{int}} = \mathbf{U}^T\mathbf{f}_{\text{int}}, \quad \tilde{\mathbf{f}}_{\text{ext}} = \mathbf{U}^T\mathbf{f}_{\text{ext}}. \tag{4}$$

ArtiSynth normally employs a lumped mass model in which $\mathbf{M}$ is constant, and so $\tilde{\mathbf{M}}$ is also constant and can be precomputed. The damping matrix $\mathbf{D}$ is also often constant and so $\tilde{\mathbf{D}}$ is also typically easy to determine. However, in any model involving large deformations, $\mathbf{K}$ is almost always nonconstant, and must be reevaluated at each simulation time step by integrating the stress/strain relationships of the model's constitutive materials over a set of integration points within each FEM model element [5]. This process, sometimes known as *matrix assembly*, has $O(n)$ complexity, with a proportionality constant depending on the nodal connectivity.

If $\tilde{\mathbf{K}}$ is evaluated using (4) directly, the resulting complexity will be $O(r^2n)$ (since $\mathbf{K}$ is sparse with $O(n)$ entries). For larger r, however, this can be burdensome. A more efficient approach is to use a smaller number of integration points, generally $O(r)$ (such that the number of points is proportional to r and not n). For example, one can select $O(r)$ elements, use a single integration point in the middle of each, and then rather than forming $\mathbf{K}$, instead accumulate the local stiffness matrix $\mathbf{K}_j$ associated with each integration point directly into $\tilde{\mathbf{K}}$:

$$\tilde{\mathbf{K}} = \sum_j \mathbf{U}^T \mathbf{K}_j \mathbf{U}.$$

Because each $\mathbf{K}_j$ has $O(1)$ size, the resulting $\tilde{\mathbf{K}}$ can be formed in $O(r^3)$ [1].

It is also possible to show that for $O(r)$ integration points, $\tilde{\mathbf{f}}_{int}$ can be determined in $O(r^2)$ time [1]. Other computations involving the reduction of external forces, and updating of the original FEM nodal positions (such as for graphic display), have a complexity of $O(nr)$.

2.2 *Application to an FEM Tongue Model*

As a test case, we applied the above reduction method to a finite element model of the human tongue [8] for modeling the tongue motions associated with speech production. The mesh is hex dominant and contains 4255 elements and 2961 nodes. The constitutive material is the same as that used for an earlier model [6]: a nearly incompressible Mooney Rivlin material with $C_{10} = 1037$, $C_{20} = 486$, and bulk modulus $\kappa = 10370$.

Tongue deformation is effected by embedding within the model a number of fiber fields corresponding to the different tongue muscles. Each field is associated with an additional anisotropic constitutive law that results in directed stresses along the fiber directions when the muscle is activated. The work described in this paper uses five tongue muscles: the hyoglossus (HG), inferior longitudinal (IL), superior longitudinal (SL), posterior genioglossus (GGP), and the transversalis (TRANS). Figure 1 shows a cutaway view of the tongue mesh, along with a representation of the fiber directions for these muscles.

Our study consisted of creating two different reduced models for this tongue, using two different basis generation techniques, and then examining how well these were able to recreate three different tongue motions (protrusion, retroflexion, and

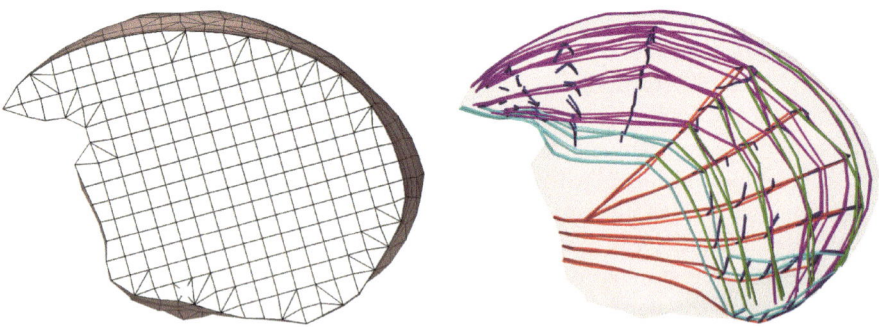

Fig. 1 Left: cutaway view of the FEM tongue model, showing the mesh structure; right: fiber directions for the five muscles (HG, IL, SL, GGP, and TRANS) used in this paper

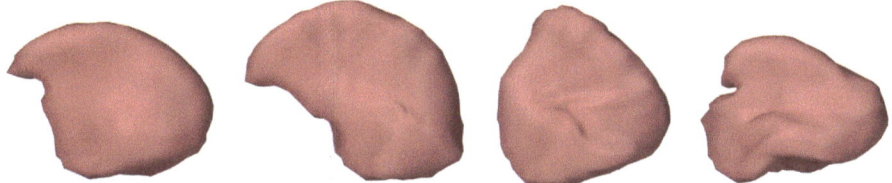

Fig. 2 Different tongue motions used in this study. From left to right: rest position, protrusion, retroflexion, and retraction

Table 1 Maximum muscle excitation levels associated with the different motions

Motion	GGP	TRANS	SL	HG	IL
Protrusion	0.4	0.4	0.1	–	–
Retroflexion	–	0.4	0.6	–	–
Retraction	–	–	–	0.6	0.4

retraction) as produced by the original model. Figure 2 shows the different motions, while Table 1 shows the muscle excitations associated with each.

The first basis was generated using modal analysis combined with a linear extension technique, as described in [22], which provides the extra degrees of freedom needed to accommodate large deformations. Specifically, six linear modes were extended by applying nine linearly independent affine transformations, resulting in a basis with $r = 54$ vectors. The second basis was determined by a training technique as described in [10], during which the tongue was exercised through activation of all muscles, one at a time, and the displacements of all nodes where recorded. The final reduced basis, consisting of 20 modes, was then determined through a principal orthogonal decomposition of the recorded data and selection of the most significant modes.

As mentioned in Sect. 2.1, the solution complexity for the reduced model is $O(r^3)$ only if the number of integration points is $O(r)$. In particular, this means that the number of integration points must be subsampled in comparison to the original model. This can be achieved in a couple of ways: (a) by choosing a certain number of points randomly within the model and (b) by using a training method to select the integration points, based on [22]. The training approach can be expected to give better results since it has more ability to ensure sufficient coverage of the muscle fiber fields, without which simulation fidelity may be lost. Results for both approaches are presented below. Also, to get a better idea of how error varied with the number of sampled integration points, we computed three different distributions of 600, 300, and 200 points, respectively.

To compare the effectiveness of the reduced model under each basis, each of the three tongue motions was executed, using 600 integration points, over a one second time interval, with the excitations ramped from zero to full strength over the interval $t \in [0, 0.5]$ s and the displacement then being allowed to settle for an additional 0.5 s.

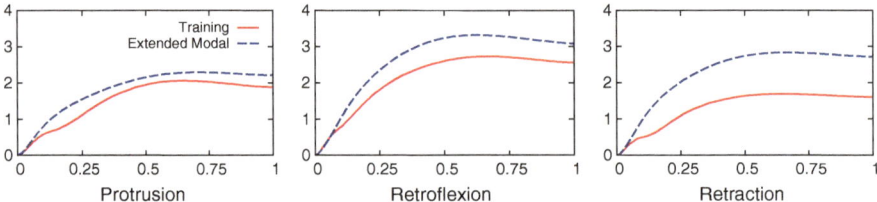

Fig. 3 Mean average errors (mm) versus time (s) for a reduced model and each of the three tasks, using both an "extended modal" and "trained" basis. The reduced model used 600 integration points. In all cases, the error was less for the trained basis

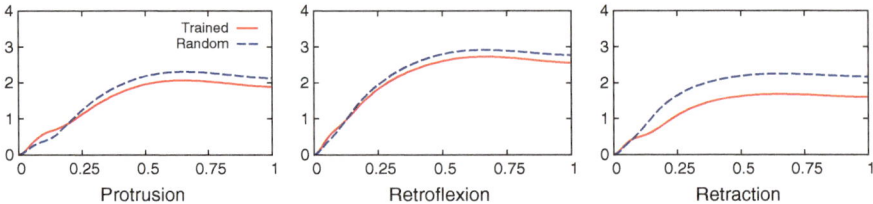

Fig. 4 Mean average errors (mm) versus time (s) for a reduced model (with a trained basis) and each of the three tasks, using 600 integration points determined by both training and random selection. In all cases the error was less for the trained integration points

These motions were then compared with those of the original (unreduced) model, with the error associated with each node's position being computed and averaged to determine a *mean average error* over time. The results for each motion are shown in Fig. 3, with mean average errors given in mm. For a sense of scale, the approximate tongue diameter is around 60 mm. These results show that the trained basis exhibited somewhat less error for a significantly smaller value of r (20 vs. 54). We also observe that the reduced model has some dynamic lag with respect to the unreduced model, as evidenced by the small error overshoot near $t = 0.5$.

To study how the error was affected by the choice of integration points, the same tests were performed again (using the trained basis), with different point distributions. First, we tested the difference between choosing points randomly versus identifying them with a training method. The results, for 600 points, are shown in Fig. 4. The training method produces slightly lower errors for two cases and significantly lower errors for another. This suggests the while training is likely to be preferred, a random method may work instead if the number of points is sufficiently large. Second, we tested the effect of using fewer points (200 and 300); results for this are shown in Fig. 5. To provide a better sense of the worst-case error, static views are shown of the unreduced and 300 point reduced model at the point of maximum error. In general, fewer integration points results in a larger error, but not always: for the protrusion task, 300 integration points give a larger error than 200. The dynamic lag also tends to increase as the number of points decreases.

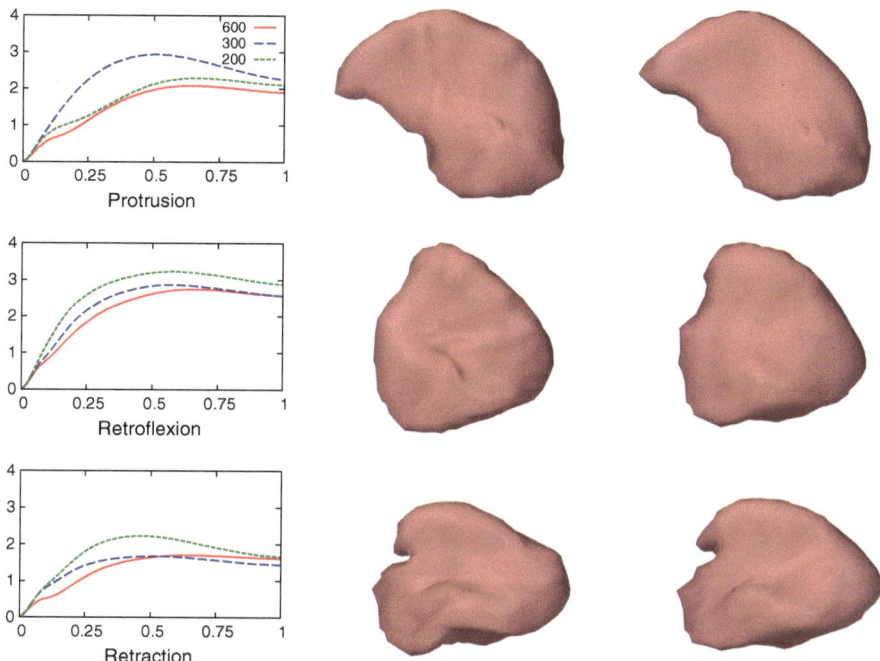

Fig. 5 Mean average errors (mm) versus time (s) for a reduced model (with a trained basis) and each of the three tasks, with the number of integration points varying between 600, 300, and 200. To get a sense of the absolute error, static images are provided of the unreduced model (middle) and 300-point reduced model (right) at the point of maximum error

Table 2 Average per-step computation times (ms) for each basis with different numbers of integration points

Basis	600	300	200
Extended modal	53.2	33.0	23.6
Trained	27.7	15.2	12.1

Reduced model effectiveness is also illustrated by an online video, showing a side-by-side comparison of task execution between an unreduced model and a reduced model with 300 integration points:

www.artisynth.org/videos/CMBBE2018ModelReduction.mp4.

Finally, average per-step computation times were compared for each of the bases, using 200, 300, and 600 integration points, with the results shown in Table 2. Computations were performed on an Intel quad core i7-7700 desktop with 16 GB of RAM. The average per-step computation time for the unreduced model was 188.9 ms, using the Pardiso [19] multicore direct solver utilizing a hybrid direct/iterative scheme. The reduced computation times were roughly linear in the number of integration points, as we would expect from the discussion of Sect. 2.1, while appearing to be roughly linear in r. However, that is because of the $O(nr)$ overheads associated with system assembly and the updating of nodal positions from the reduced coordinates.

Typical maximum absolute errors in the above results are around 3 mm, which for an approximate tongue diameter of 60 mm corresponds to a relative error of around 5%. Meanwhile, for 300 or 200 integration points, the speed-up with the trained basis is more than 10 times, and reduces the average per-step computation time to interactive rates.

2.3   *Application to an FEM Foot Model*

As an example involving a larger FEM model containing internal rigid structures, we describe preliminary work on model reduction for the detailed biomechanical foot model described in [15]. Developed in ArtiSynth, this contains all the bones of the foot, together with tendons and ligaments, embedded within an FEM model that emulates the skin and other soft tissues (Fig. 6). The bones are modeled as rigid bodies, to which nearby FEM nodes are connected using point-to-frame attachments. The FEM model itself uses a neo-Hookean material and contains 23,298 elements and 13,087 nodes. Joints between the bones are modeled using unilateral contact, which provides more realistic joint motions, but, when combined with the FEM embedding, increases per-step computation times to around 2600 ms on a an Intel quad-core i7-7700 desktop with 16 GB of RAM. When run by itself without any embedded structures, the FEM model has an execution speed of around 650 ms.

To help make this model suitable for clinical applications, we are investigating model reduction to reduce its computation time to the point where it can be run interactively. We developed a reduced basis for the model by creating a series of nodal displacements corresponding to various excitement levels of the tibialis anterior muscle and then applying a principal orthogonal decomposition in a manner similar to that used for the tongue model (Sect. 2.2). The basis contains 6 vectors, and when combined with 100 randomly chosen integration points allows the model to be run at a real-time rate of around 10 ms/step (Fig. 7). Further investigations will consider other muscle excitations and external loadings such as floor contact.

Fig. 6 The foot model, showing bones, tendons, ligaments, and the embedding FEM mesh

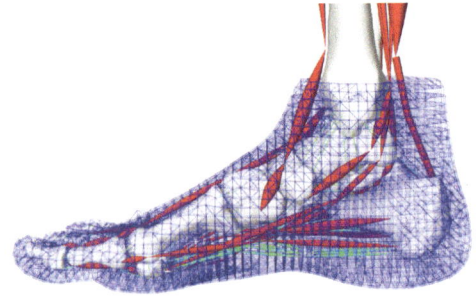

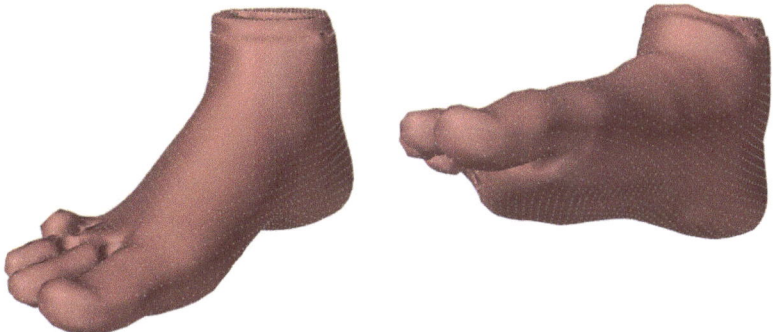

Fig. 7 The reduced foot model at rest (left) and after muscle excitation (right)

3 Attaching Points and Frames to Deformable Bodies

Biomechanical models are often built using a variety of components and sub-models, which must then be connected together. An illustrative example of this is the FRANK model [2] (Fig. 8), which provides a reference model of human head and neck anatomy. Implemented in ArtiSynth [12], FRANK is designed to simulate anatomical functions related to swallowing, chewing and speech, and consists of various components modeled using finite elements, rigid bodies, point-to-point springs, and muscles structures. ArtiSynth provides a number of means for connecting such components together, including general constraints, joints, and the ability to directly attach points and coordinate frames directly to other components. This section focuses on the latter capability, and describes the recently enhanced mechanism for connecting either points or frames to deformable bodies. This allows, for example, a point-to-point muscle to be connected directly to an FEM tissue model.

The ArtiSynth attachment mechanism works by defining the coordinates $\mathbf{x}_a$ of the *attached* component to be a function of the coordinates $\mathbf{x}_m$ of one or more *master* components to which it is attached:

$$\mathbf{x}_a = f(\mathbf{x}_m). \tag{5}$$

This then implies that the velocities are related by a linear relationship of the form

$$\dot{\mathbf{x}}_a = \mathbf{G}_{am}\dot{\mathbf{x}}_m, \quad \mathbf{G}_{am} \equiv \nabla f(\mathbf{x}_m). \tag{6}$$

From the principle of virtual work, discussed above, forces $\mathbf{f}_a$ on the attached components then propagate back to forces $\mathbf{f}_m$ on the master components via

$$\mathbf{f}_m = \mathbf{G}_{am}^T \mathbf{f}_a. \tag{7}$$

Fig. 8 The FRANK model of human head and neck anatomy, with some structures (such as the jaw) not shown to reveal internal detail

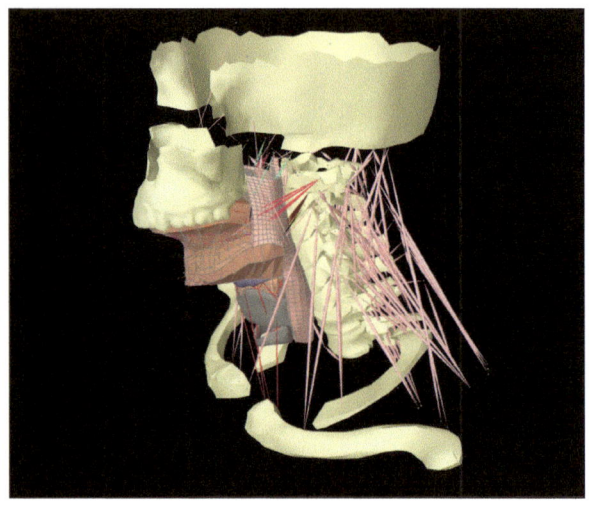

3.1 Point Attachments

For the case of a point attached to an FEM model, its position (and velocity) is given by a weighted sum of nearby nodal positions $\mathbf{x}_j$:

$$\mathbf{x}_a = \sum_j w_j \mathbf{x}_j, \quad \dot{\mathbf{x}}_a = \sum_j w_j \dot{\mathbf{x}}_j.$$

Forces $\mathbf{f}_a$ applied to the point then propagate back to each node according to

$$\mathbf{f}_j = w_j \mathbf{f}_a.$$

Often the local nodes are chosen to be the ones associated with the element containing the node, but in some cases, it may be desirable to spread the attachment across a larger set of nodes, in order to better distribute forces imparted by the attached point across the FEM model (Fig. 9). A good case example of this is shown in Fig. 10, where the styloglossus muscles of a tongue are modeled as external point-to-point muscles connected to the main FEM tongue model.

Points can be attached to a reduced model in essentially the same way, only now the support nodes are themselves controlled by the underlying reduced coordinates:

$$\mathbf{x}_a = \sum_j w_j (\mathbf{x}_{j0} + \mathbf{U}_j \mathbf{q}), \quad \dot{\mathbf{x}}_a = \sum_j w_j \mathbf{U}_j \dot{\mathbf{q}},$$

where $\mathbf{x}_{j0}$ and $\mathbf{U}_j$ are the rest position and the submatrix of $\mathbf{U}$ corresponding to node j. With respect to (6), $\mathbf{G}_{am}$ takes the form

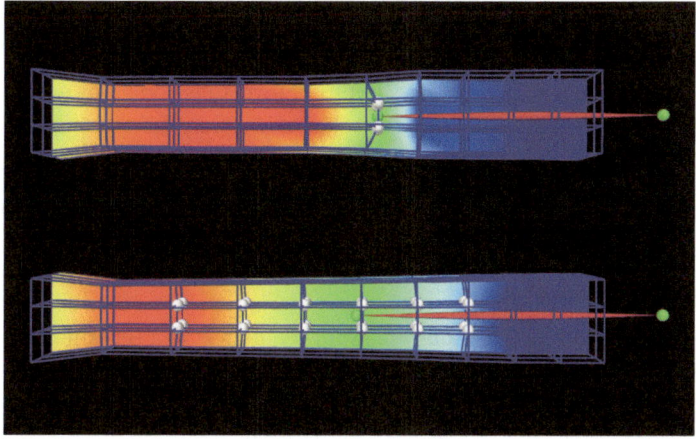

Fig. 9 Two examples of a point-to-point muscle attached to an FEM model, using 4 support nodes (top) and 24 support nodes (bottom). The resulting stress/strain pattern is smoother and more diffuse with the larger number of support nodes

Fig. 10 A finite element model of the tongue, with the styloglossus modeled as externally connected point-to-point muscles (red). It may be desirable to distribute the styloglossus/tongue attachment across the nodes of multiple elements

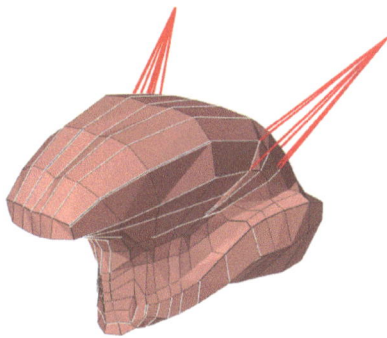

$$\mathbf{G}_{am} = \sum_j w_j \mathbf{U}_j.$$

One difference between reduced and FEM model attachments is that for the former it is often less necessary to be concerned about distributing the stress/strain across multiple nodes, since the model reduction process tends to do this automatically.

A current open problem is the ability to directly connect two reduced models together. This is because there is no easy way to guarantee that the reduced motion of each body would be mutually compatible with the attachment, particularly when the attachment has spatial extent. Any future implementation of such an attachment would presumably need to "relax" its constraints to accommodate the motion range of the reduced models.

3.2   Frame Attachments

Coordinate frames can also be connected to deformable bodies in much the same way as for points. The frame origin is attached as a point, and the orientation **R** can then be determined in one of two ways:

- *Element shape functions*: If the nodes are associated with an element, then the local deformation gradient **F** can be determined using element shape functions in the standard FEM manner, with **R** then determined from **F** using a polar decomposition **F** = **RP**.
- *Procrustean method*: If the nodes are arbitrary, then **R** can be estimated based on a Procrustean approach. First we compute a matrix **F** according to

$$\mathbf{F} = \sum_j w_j (\mathbf{r}_j \mathbf{r}_{0j}^T),$$

where w_j are the nodal weights and $\mathbf{r}_j$ and $\mathbf{r}_{0j}$ are the current and rest positions of the nodes described with respect to the coordinate frame origin. **R** can then again be determined from **F** using a polar decomposition **F** = **RP**.

The ability to connect frames means in particular that rigid bodies can be connected directly to deformable models, as shown in Fig. 11, right.

Since frames can be attached to deformable bodies, this also means that joints (which implement constraints between frames) can also be attached to deformable bodies. A useful application of this is the subject-specific registration of skeletal anatomy, as presented in [17] and illustrated in Fig. 12. This involves taking a reference model of a skeletal structure, including bones and joints, and registering (i.e., deforming) it to a specific subject based on medical imaging data (e.g., an CT scan). To do this, the bones in the reference model must be made deformable (which can be done by placing them within an embedding mesh, as described in Sect. 4) and then "attracted" to the subject data using a technique such as iterative closest point

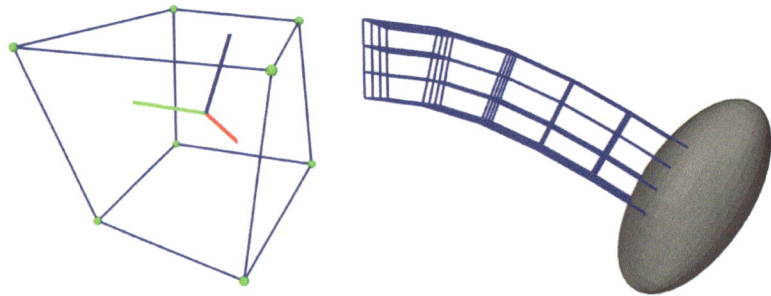

Fig. 11 Attaching frames to deformable bodies. Left: a frame connected directly to the nodes of a single FEM element. Right: an ellipsoidal rigid body connected to the end of an FEM beam

Fig. 12 Subject-specific registration of skeletal structures. Each bone mesh of a reference model (top) is made deformable by embedding it within a regular FEM grid (shown here for the ulna). The deformable bones are then connected using FEM joints, allowing the reference to be more easily registered to subject-specific data (bottom) using ICP or similar techniques

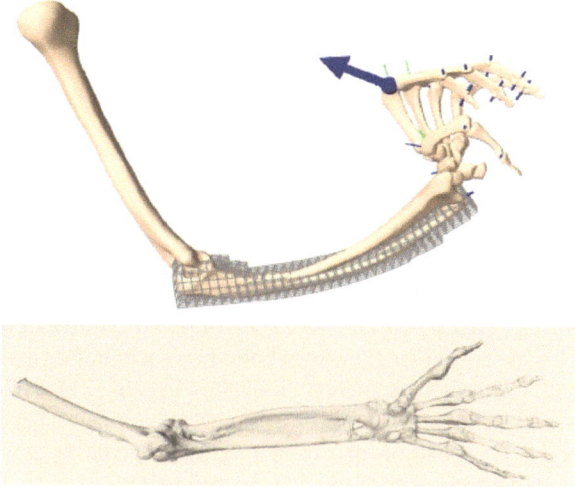

(ICP) [4]. In order to preserve the structural integrity of the reference model, the (deformable) bones are connected with joints appropriate to the anatomy, allowing the model to bend freely at each joint (up to joint limits) and to simultaneously register the shape and pose of each bone. If instead the reference was modeled as a single deformable structure without such joints, it would be necessary to greatly reduce the stiffness near each joint location, in an anisotropic way that emulated the constraints of each joint.

4 Skinning and Embedded Meshes

Another useful technique for creating anatomical and biomechanical models are to attach a passive mesh to an underlying set of dynamically active bodies so that it deforms in accordance with the motion of those bodies. ArtiSynth allows meshes to be attached to collections of both rigid bodies and FEM models, facilitating the creation of structures that are either embedded-within, connected-to, or enveloped-by a set of underlying components. Such mesh embedding approaches are well known in the computer graphics community and have more recently been applied to biomechanics [13], and also figure prominently in the SOFA system [7].

It should be noted that mesh embedding provides an alternate way to perform model reduction, in the sense that the number of dynamic DOFs for the resulting system is determined by the number of nodes in the embedding mesh. By using a course embedding mesh, the number DOFs can be significantly reduced.

The underlying method uses the attachment mechanism (Eqs. (5)–(7)), with the mesh vertices being the "attached components". The mesh deforms in response to the attachment configuration, while external forces applied to the mesh can be propagated back to the dynamic components via (7). This technique is useful in a variety of applications, as presented below.

4.1 Skinning for Modeling the Human Airway

One application is to create a continuous skin surrounding an underlying set of anatomical components. For example, for modeling the human airway, a disparate set of models describing the tongue, jaw, palate and pharynx can be connected together with a surface skin to form a seamless airtight mesh (Fig. 13), as described in [21]. This then provides a uniform boundary for handling air or fluid interactions associated with tasks such as speech or swallowing. In [21], each skin mesh vertex is attached to one or more *master* components, which can be either an FEM model or a rigid body. The position $\mathbf{q}_v$ of each vertex is given by a weighted sum of contributions from the master components, according to

$$\mathbf{q}_v = \mathbf{q}_{v0} + \sum_i^M w_i f_i(\mathbf{q}_m, \mathbf{q}_{m0}, \mathbf{q}_{v0}) \tag{8}$$

where $\mathbf{q}_{v0}$ is the initial position of the skinned vertex, $\mathbf{q}_m$ and $\mathbf{q}_{m0}$ give the positions and rest positions of the M master components, and w_i and f_i are the skinning weight and blend function associated with the ith master component. Further details are given in [21]. The position equation (8) can be differentiated to yield a velocity relationship

$$\mathbf{u}_v = \mathbf{G}\mathbf{u}_m,$$

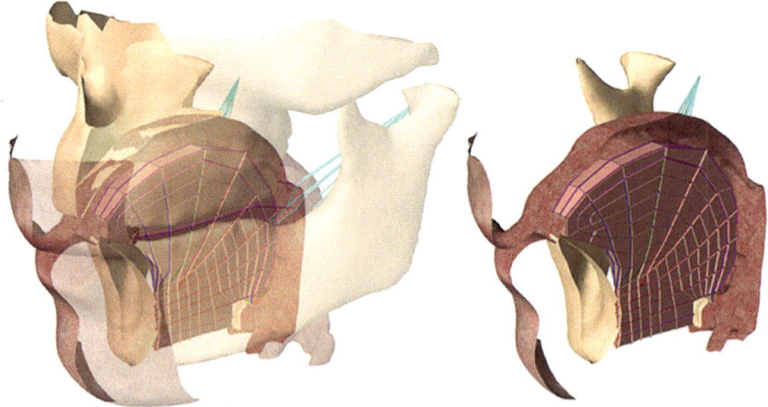

Fig. 13 A skin mesh used to delimit the boundary of the human upper airway, connected to various surrounding structures including the palate, tongue, and jaw [21]

where $\mathbf{u}_v$ and $\mathbf{u}_m$ are the velocities of the skin vertex and the master components and $\mathbf{G}$ is a (local) matrix. Then from the principle of virtual work, vertex forces $\mathbf{f}_v$ can be propagated back onto the master component forces $\mathbf{f}_m$ via

$$\mathbf{f}_m = \mathbf{G}^T \mathbf{f}_v.$$

In general, this means that forces, pressures, or contact interactions applied to the skin can be reflected back onto the underlying master components. In the case of the airway model, these external loads could involve pressures from air, fluid, or food bolus interactions.

4.2 Mesh Embedding Applied to Modeling the Masseter

Embedding and attachment techniques have recently been applied to the creation of a finite element model of the human masseter [16], which is the principal muscle used in chewing (Fig. 14). This model contains detailed information about the internal muscle fascicles and aponeuroses, with a primary purpose being to study the importance of aponeuroses stiffness in the transmission of force within the masseter.

Fascicle and aponeuroses data were obtained using the dissection and digitization procedure of [9]. Fascicle data was discretized into a set of line segment meshes, while the aponeuroses (tendon sheets) were discretized into triangular surface meshes. A muscle volume was then constructed by constructing a wrapping surface around this fiber and aponeuroses data, in a manner similar to [11]. Originally, this was used as the surface mesh from which a conforming hex-dominant volumetric mesh was constructed [18] (Fig. 15, middle). However, creating such a conforming mesh was both time consuming and also yielded a number of poorly conditioned elements, and so a mesh embedding approach was adopted instead in which the muscle volume surface was embedded inside a coarse but highly regular and well-conditioned grid (Fig. 15, right). Model fidelity was improved using the technique of [13], in which the mass and stiffness values of the embedding FEM are weighted to account for regions where the embedded mesh is absent.

Fig. 14 Model of the masseter connected to the jaw and mandible

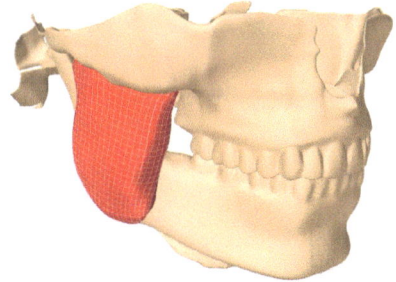

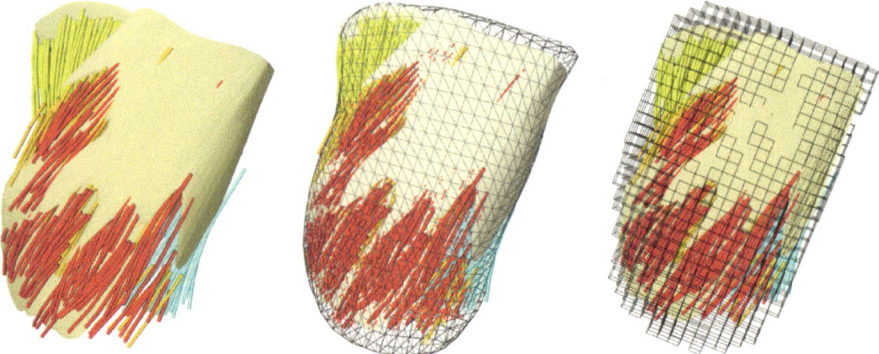

Fig. 15 Left: raw digitized data for masseter fascicles and aponeuroses. Middle: FEM masseter model based on a conforming mesh generated from a wrapping surface enclosing the data. Right: FEM model based on a regular embedding mesh. For both models, fascicles and aponeuroses are added as embedded structures

Fascicles and aponeuroses were also incorporated within this primary FEM mesh. Fascicle data was embedded within the primary mesh, hence allowing it to deform accordingly. The fiber directions were used to determine the direction of muscle contraction used by the muscle constitutive law at nearby integration points of the primary mesh. Aponeuroses were modeled as thin membrane-like FEM models, created by extruding their triangular surface data using wedge elements. These membrane FEMs were then attached to the primary FEM by connecting each membrane node to its containing element within the primary mesh (as described in Sect. 3.1), allowing the membrane stiffness to be transferred onto the entire structure.

By using embedding and attachment techniques, it is possible to create a masseter model with far fewer degrees of freedom (and hence a much faster simulation time) than would otherwise be possible. The embedding technique allows the primary mesh to be set at a resolution appropriate to the overall deformability of the muscle, rather than a need to accommodate the surface structure. More importantly, the use of attachments to connect the aponeuroses allow the primary and aponeuroses meshes to be nonconforming. Otherwise, it would be necessary to employ meshes with far higher resolutions, which would be both harder to construct and would result in much higher simulation times.

5   Conclusion

We have described a number of useful methods for enhancing the construction of biomechanical models. The first, model reduction, allows applications to implement complex deformable models at reasonable computational cost, and we are currently in the process of introducing this into the ArtiSynth modeling system. Most of the effort associated with model reduction involves determining both the basis and integration point distribution. Our results suggest that training techniques tend to yield the best basis results, yielding both a smaller basis and less error. Training is also useful for selecting integration points, although random point selection may sometimes work sufficiently well. We have also seen that model reduction has the potential to improve computational speeds to interactive rates.

The other methods include ways to attach points and frames to finite element models, along with skinning and mesh embedding. All of these are currently available in ArtiSynth and facilitate the dynamic interconnection of model components and the introduction of auxiliary mesh structures for both visualization and simulation purposes. They have already been applied to a number of applications in biomechanics; those described here include a large scale reference model of the head and neck, subject-specific skeletal registration, skinning applied to modeling the human airway, and a detailed model of the human masseter.

The unifying concept underlying all these techniques is the principle of virtual work, which explains the force relationship between coordinate systems when the velocity relationship is known.

References

1. An SS, Kim T, James DL (2008) Optimizing cubature for efficient integration of subspace deformations. ACM Trans Graph (TOG) 27(5):165
2. Anderson P, Fels S, Harandi NM, Ho A, Moisik S, Sánchez CA, Stavness I, Tang K (2017) Frank: s hybrid 3d biomechanical model of the head and neck. Biomechanics of living organs. Elsevier, Amsterdam, pp 413–447
3. Barbič J, James DL (2005) Real-time subspace integration for st. venant-kirchhoff deformable models. ACM Trans Graph (TOG) 24(3):982–990
4. Besl PJ, McKay ND (1992) Method for registration of 3-d shapes. In: Sensor fusion iv: control paradigms and data structures, vol 1611. International Society for Optics and Photonics, pp 586–607
5. Bonet J, Wood RD (2008) Nonlinear continuum mechanics for finite element analysis. Cambridge University Press, Cambridge
6. Buchaillard Stéphanie, Perrier Pascal, Payan Yohan (2009) A biomechanical model of cardinal vowel production: muscle activations and the impact of gravity on tongue positioning. J Acoust Soc Am 126(4):2033–2051
7. Faure F, Duriez C, Delingette H, Allard J, Gilles B, Marchesseau S, Talbot H, Courtecuisse H, Bousquet G, Peterlik I et al (2012) Sofa: a multi-model framework for interactive physical simulation. Soft tissue biomechanical modeling for computer assisted surgery. Springer, Berlin, pp 283–321
8. Hermant N, Perrier P, Payan Y (2017) Human tongue biomechanical modeling. Biomechanics of living organs. Elsevier, Amsterdam, pp 395–411
9. Kim SY, Boynton EI, Ravichandiran K, Fung LY, Bleakney R, Agur AM (2007) Three-dimensional study of the musculotendinous architecture of supraspinatus and its functional correlations. Clin Anat Off J Am Assoc Clin Anat Br Assoc Clin Anat 20(6):648–655
10. Krysl P, Lall S, Marsden JE (2001) Dimensional model reduction in non-linear finite element dynamics of solids and structures. Int J Numer Methods Eng 51(4):479–504

11. Lee Dongwoon, Ravichandiran Kajeandra, Jackson Ken, Fiume Eugene, Agur Anne (2012) Robust estimation of physiological cross-sectional area and geometric reconstruction for human skeletal muscle. J Biomech 45(8):1507–1513
12. Lloyd JE, Stavness I, Fels S (2012) Artisynth: a fast interactive biomechanical modeling toolkit combining multibody and finite element simulation. Soft tissue biomechanical modeling for computer assisted surgery. Springer, Berlin, pp 355–394
13. Nesme M, Kry PG, Jeřábková L, Faure F (2009) Preserving topology and elasticity for embedded deformable models. ACM Trans Graph (TOG) 28:52 ACM
14. Pentland A, Williams J (1989) Good vibrations: modal dynamics for graphics and animation. Comput Graph 23(3):207–214
15. Perrier A, Luboz V, Bucki M, Cannard F, Vuillerme N, Payan Y (2017) Biomechanical modeling of the foot. Hyperelastic constitutive laws for finite element modeling. Biomechanics of living organs. Academic Press, Cambridge, pp 545–563
16. Sánchez CA, Li Z, Hannam AG, Abolmaesumi P, Agur A, Fels S (2017) Constructing detailed subject-specific models of the human masseter. Imaging for patient-customized simulations and systems for point-of-care ultrasound. Springer, Berlin, pp 52–60
17. Sanchez CA, Lloyd JE, Li Z, Fels S (2015) Subject-specific modelling of articulated anatomy using finite element models. In: Proceedings of 13th international symposium on computer methods in biomechanics and biomedical engineering (CMBBE)
18. Sanchez CA, Stavness I, Lloyd JE, Hannam AG, Fels S (2014) Modelling mastication: the important role of tendon in force transmission. In: Proceedings of 12th international symposium on computer methods in biomechanics and biomedical engineering (CMBBE)
19. Schenk Olaf, Gärtner Klaus (2004) Solving unsymmetric sparse systems of linear equations with pardiso. Futur Gener Comput Syst 20(3):475–487
20. Sifakis E, Barbic J (2012) FEM simulation of 3d deformable solids: a practitioner's guide to theory, discretization and model reduction. In: Proceedings of ACM SIGGRAPH 2012 courses. ACM, p 20
21. Stavness I, Sánchez CA, Lloyd J, Ho A, Wang J, Fels S, Huang D (2014) Unified skinning of rigid and deformable models for anatomical simulations. In: Proceedings of SIGGRAPH Asia 2014 technical briefs. ACM, p 9
22. von Tycowicz C, Schulz C, Seidel H-P, Hildebrandt K (2013) An efficient construction of reduced deformable objects. ACM Trans Graph 32(6):213:1–213:10

Computational Cell-Based Modeling and Visualization of Cancer Development and Progression

Jiao Chen, Daphne Weihs and Fred J. Vermolen

Abstract This paper presents a review of the role of mathematical modeling in investigating cancer progression, focusing on five models developed in our group. A brief overview of computational modeling progress is presented, followed by introduction of several mathematical formalisms (e.g., stochastic differential equations), numerical methods (e.g., finite element method, Green's functions, and combinations of time integration), and Monte Carlo simulations, which are currently used to quantify the underlying biomedical mechanisms, to approximate the results and to evaluate the impact of the input variables. Next, we provide specific examples of the computational models that we developed aimed at predicting the dynamics of the initiation and progression of cancer. Our simulation results show qualitative consistency with references and/or available experimental observations. Finally, perspectives are drawn on the possibilities of mathematical modeling for the prospects of cancer understanding and treatment therapies.

Keywords Mathematical modeling · Numerical method · Cancer progression · Cell migration · Angiogenesis · Metastasis · Immune responses · Cell deformation

1 Introduction

Cancer has become one of the leading causes of death in developed countries and its global mortality rate is rising [55]. Cancer initiates and develops by a series of processes comprising cell mutation, abnormal proliferation, angiogenesis, and

J. Chen (✉) · F. J. Vermolen
Delft Institute of Applied Mathematics, Delft University of Technology, Delft, The Netherlands
e-mail: j.chen-6@tudelft.nl

F. J. Vermolen
e-mail: F.J.Vermolen@tudelft.nl

D. Weihs
Faculty of Biomedical Engineering, Technion-Israel Institute of Technology, Haifa, Israel
e-mail: daphnew@technion.ac.il

© Springer Nature Switzerland AG 2019
J. M. R. S. Tavares and P. R. Fernandes (eds.), *New Developments on Computational Methods and Imaging in Biomechanics and Biomedical Engineering*, Lecture Notes in Computational Vision and Biomechanics 33, https://doi.org/10.1007/978-3-030-23073-9_7

metastasis accompanied by the evolution of cell morphology. Many studies have utilized in vitro systems using primary cells or cell lines. While those studies have provided an important understanding of cancer, the interaction between cancer cells and their microenvironments are difficult to reproduce accurately. Thus, for more physiologically relevant conditions, animal experiments have been used.

Animal-based experiments have been crucially important in cancer research, in particular, in cancer pathology, tumor transplant, immunization, and treatment. However, the cruelty and ethical views caused by animal experiments have caused a reduction in their use. In 1959, the conception of "Three Rs" was proposed as the principles of Replacement, Reduction, and Refinement in *The Principles of Humane Experimental Technique* and the concept has been a hot issue in the EU legislation since 1986, aimed at protecting animals [26, 90]. Therefore, in the wake of research requirements and experimental regulations, well-designed experiments are crucially important, which definitely need the input from various disciplines like mathematics, physics, computer science, etc.

To validate developed hypotheses regarding biological processes occurring during cancer development, it is necessary to assess experimental outcomes. Since experimental results are usually represented in the form of patterns and numbers, the developed hypotheses need quantification. This quantification requires the translation of hypotheses into quantitative relations, which generally pose a set of mathematical relations. The combination of the mathematical relations constitutes the backbone of the mathematical model that is used to simulate the biological process of interest. Mathematical models are capable of reproducing situations that are beyond the measured data. Despite all advantages of mathematical modeling, one should be careful in the evaluation of the results due to possible shortcomings in the model. Shortcomings in mathematical models arise from neglecting several features in biological processes due to lack of knowledge, as well as by uncertainty of parametric values. The latter shortcoming requires parameter sensitivity analysis. Facing the societal burden, mathematical modeling of cancer is a promising approach to combine with experiments in *vitro* and in *vivo*, using both animal and human materials. On one hand, the modeling results lead to predictions [45] and further description with examples can be found in Gammon [44]. On the other hand, with computational modeling, the number of animal trials could be reduced and the experiments can be designed better, however, conversely, mathematical models could be validated by corresponding experiments.

As early as in 1942, a book named "On Growth and Form" by Thompson et al. [110] cited the following quote from a statistician Karl Pearson (first published in 1901):

> I believe the day must come when the biologist will - without being a mathematician - not hesitate to use mathematical analysis when he requires it

and presents mathematical principles in his book. A 100 years later, a paper in 'The Economist' (2004) stated that

> If cancer is ever to be understood properly, mathematical models such as these will surely play a prominent role.

1.1 Mathematical Modeling on Various Scales

Mathematical models have been developed for a broad spectrum of length scales, ranging from a molecular level (from a few atoms to multitudes of biomolecules) to a tissue level. Fearon and Vogelstein [38] describe a conceptual model showing cancer evolution as a series of genetic mutation mainly in tumor oncogenes and suppressor genes. With a further rigorous verification, Gatenby and Vincent [46, 124] find that environmental selection forces are dominated by competition for limited substrate in the era of carcinogenesis. Next to genes, due to a large amount of proteins involved in cancer development and progression, where some of them even become the targets of new drugs, molecular modeling is important and is able to provide details that would not be accessible if solely experiments on the molecular dynamics [43] were carried out. For example, Wonpil et al. [53] proposes Brownian dynamics for modeling the movement of ions in membrane channels. Binding of proteins and DNA have been described by Chuanying Chen and Pettitt [19]. Furthermore, molecular mechanics have been described by Spiegel and Magistrato [106], Turjanski et al. [112], and other molecular models refer to a review paper by Friedman et al. [43].

Cells constitute the fundamental, independent functional unit of organisms. Cancerous cells display many features compared to the characteristics of normal constitutive cells, which have been incorporated in various mathematical formalisms. In a cell-based modeling framework, the geometry of one cell can be fixed, see for instance, [10, 13, 18, 22, 31, 123]; whereas, the cell morphology is also changeable in reality. Rejniak [93], Rejniak and Dillon [95] utilize an immersed boundary approach with distributed sources to model the deformable boundary of cells at early stages with an application to ductal tumor. Moreover, deformation of cells can also be realized through the simulation of cytoskeleton [120]. Furthermore, the studies by Madzvamuse [70], Elliott et al. [34] treat the evolving geometry of the cell membrane by combining a moving boundary problem with a system of coupled surface partial differential equations, which are solved by the use of surface finite element methods.

For the study of cancer progression and disease pathology, the modeling of large ensembles of interacting cells in biological tissue is needed. A literature study by Murray [81] proposes several partial differential equation-based models to simulate various biological phenomena like wound healing, cancer development, and immune system response on the macro tissue scale. In the context of cancer dissemination and metastasis, clusters of cells have much higher metastatic potentials than singular migrating cancer cells [58, 76]. Based on this, Dudaie et al. [32] describe a model on the collective movement of cancer cells on a cell colony level and Jolly et al. [58] developed a model for investigating cluster-based dissemination of breast cancer cells. Taking the CPU time into consideration, parallel computing platforms are feasible for tissue simulation involving large numbers of interacting cells [28].

1.2 Mathematical Modeling from a View of Cancer Development and Progression

Cancer development involves a chain of biophysical processes including initiation, angiogenesis, metastasis and colonization, of which some of them are sketched in Fig. 1.

With a series of gene mutations, normal cells mutate into cancer cells and obtain abnormal properties (i.e., dysfunctional excessive proliferation) [41]. According to the studies by NHS (National Cancer Intelligence Network), early diagnosis of cancer can increase the likelihood of survival significantly. However, during this period, cancer is usually difficult to detect. A literature review on cell-based models in which the initial stages of cancer have been modeled in Chen and Vermolen [20]. The importance of early diagnosis motivates the need for strengthening scientific research on the early stages of cancer development both from clinical and in-silico studies. To simulate cancer initiation, Vermolen et al. [123] develop various cell-based models and Chen et al. [22] introduce a model to describe the antitumor immune responses by tumor-specific T-lymphocytes at tumor early stage with applications to pancreatic cancer. Furthermore, Enderling et al. [36] develop a model on radiotherapy strategies targeting cancer during its early stages.

Angiogenesis triggers a key transition for tumors from being a dormant avascular phase to reaching a soaring vascular phase [40] see Fig. 1. Based on this phenomenon, many experimental studies aim at preventing angiogenesis or at cutting off oxygen sources to prevent further development and more importantly metastasis of a tumor. For example, a drug named Avastin is regarded as a powerful means for cutting off the tumor's oxygen supply, however, tumors become more aggressive as a result of

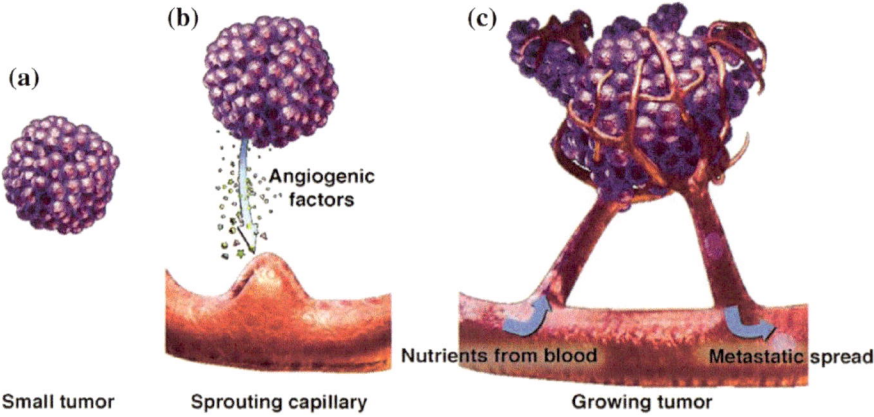

Fig. 1 The transition from a benign tumor, see left, to a malignant tumor, see right. The interaction of tumor growth and angiogenesis. Taken from Siemann DW, Vascular targeting agents. Horizons in Cancer Therapeutics: From Bench to Bedside. 2002; 3 (2): 4–15 [104]

further hypoxia [101]. To aid the experiments, mathematical modeling has yielded several contributions such as an 3D angiogenesis model [103], a cellular automata model of growth of blood vessels by Rens et al. [96], and other relevant works [17, 73, 74]. An excellent review about tumor-induced angiogenesis is written by Stephanou et al. [107].

Metastasis is responsible for as much as 90% of cancer-caused mortality [15]. In this case, the migration of cells often proceeds through mechanotaxis, which can be classified into tensotaxis (movement according to mechanical tensions) and durotaxis (migration toward a stiffness gradient). In tensotaxis migration, cells exert forces on their extracellular matrix environment, and in turn the stresses, displacements, and strains due to cell-induced deformation of the microenvironment can be sensed by neighboring cells. Those neighboring cells are then able to migrate according to the mechanical signals. This is experimentally evidenced and modeled in Reinhart–King et al. [91], Vermolen and Gefen [118], Quinlan et al. [87], Dembo and Wang [29], DuFort et al. [33]. For durotaxis migration, cells will tend to move in the direction of a stiffness gradient, and especially cancerous cells show a preference for a stiffer substrate or extracellular matrix (ECM) [33, 69, 83, 129]. The works by Weihs et al. have also shown that the stiffness of the substrate affects the ability of cancer cells to exert forces related to adherence [72] and to mechanical invasiveness [65]. Regarding cancer metastasis, some existing mathematical models can be found in the works [2, 5, 89, 99, 113].

The existence of the complexity and heterogeneity in various cancers poses a big challenge to adequate treatments. For many decades, many studies have been devoted to finding a breakthrough for cancer treatment. With the bound of ethics as well as the increasingly loud voice of anti-animal experiments, biological experiments and clinical trials need to be closely integrated with mathematics and high-speed development of computer technology. In this fast pace of life, more and more people are suffering from chronic or emerging diseases and a majority of cancers develop as a result of chronic inflammation [111]. Mathematical models provide an avenue to explore possible improved and alternative therapies against cancer [105]. For instance, Enderling et al. [35–37] model radiotherapy of breast cancer and their work showed the possibility of investigating clinically verifiable hypotheses for the influence of radiotherapy on cancer progression through numerical simulation. Furthermore, Tanaka et al. [109] propose a model for prostate cancer which is helpful to scrutinize the application of hormone therapy. Another modeling work, treating the influence of chemotherapy on cancer cells, is reported in [82]. Next to the traditional treatment approaches like surgery, chemotherapy, radiotherapy, and cancer immunotherapy has shown some prospects [27, 75]. Therefore, the numerical simulation of immunotherapy has become a research direction. Through boosting the immune system of individuals to fight cancers, a model in terms of tumor–immune interaction is developed in [64]. A survey of several mathematical models and methods dealing with the tumor–immune interaction is provided in [7]. Moreover, any process of cancer progression could be a therapeutic target and several therapy-related models are introduced by Abbott and Michor [1]. Furthermore, smart health care has drawn a lot of attention, which is able to monitor a patient's vital organs and to guide doctors to perform surgery as well as to apply medications more accurately. However, smart health care is non-separable from computer technology with numerical simulation and it faces many mathematical challenges that need to be solved. Therefore, mathematical modeling is a promising means to reduce the cost and ethical burden of experimental tests for cancer research [105] and will even contribute to quantify the impact of therapies against cancer. This quantification will be used to improve and even better, to optimize certain therapies by computing the impact of new therapies against cancer.

1.3 Mathematical Modeling from a View of Identified Cancer Types

There are currently more than a 100 distinct types of identified cancers [48]. According to the global cancer statistics given by Ferlay et al. [39], the cancer with the highest mortality rate, as high as 19.4% in adults, is lung cancer. Regarding the models on lung cancer, Chmielecki et al. [24] use an evolutionary model to optimize the dosing of drug treatment. Wang et al. [126] and Bianconi et al. [8] propose further mathematical models to simulate lung cancer by multiscale agent-based and systems biology inspired formalisms for large tumors. Furthermore, other types of cancers are simulated mathematically like liver cancer [128], breast cancer [35–37, 58], brain cancer [52, 60, 86], avascular cancer [127], prostate cancer [54, 109], etc. Some tumors develop in distinct architectural forms, e.g., preinvasive intraductal tumors in the breast or prostate, which are simulated by Rejniak and Dillon [95]. Pancreatic cancer is notorious for its profuse stroma with less than 4% 5-years survival rate [56], which is modeled as far as we know for the first time by Chen et al. [22].

1.4 Mathematical Modeling from a View of Model Types

Modeling mechanics of cancer cells and tissue is an emerging field with a broad spectrum of patterns. Agent-based models are developed to understand the mutual interplay of an individual cell and its surroundings on the microscale, compared to the macro-scale off-agent models. The macro-scale models consider rather than individual cells. Their merit is their applicability to larger physical tissue regions. For a literary review on agent-based modeling, we refer to [115], which describes three types agent-based models.

1. Lattice-based models can be further classified into various types such as cellular automata [9, 88], lattice gas cellular [98], and cellular Potts models [47, 77, 116]. Cellular automata, viewed as an approach to model complex systems, is applied to tumor invasion in [50]. In terms of tumor development, a critical review [80]

evaluates classical cellular automaton models. To unravel the potential effects of movement and interaction of individuals (like cells collisions), lattice gas cellular automata is slightly improved from classical cellular automata. The model in [51] is an example with application to cell migration. Using cellular Potts models, it is possible to define the cell more precisely. The possibilities to apply them to tumor evolution are described by Szabó and Merks [108].

2. Off-lattice models, see [57] as an example, can stimulate tumor growth and invasion. Jeon et al. [57] state that an off-lattice model enables them to model cell motility with detailed forces and to overcome the drawbacks of a lattice model (e.g., a limited set of possible directions and discrete displacements). Schaller and Meyer–Hermann [102] introduce a three-dimensional off-lattice Voronoi–Delaunay model to study multicellular tumor spheroids. Moreover, there are alternative off-lattice models like particle models, such as [14, 30, 117, 121] and cell shape evolving models such as [11, 120].

3. Hybrid discrete-continuum models are feasible for large systems [63, 78, 115, 131]. Here the hybrid model serves as an integrated method to predict cancer cell behavior in a microenvironment in terms of both discrete and continuous variables [94]. Another example of a hybrid model of interaction between tumor and stromal environment is developed for breast cancer [62].

Due to the overwhelming complexity in cancer research, the joint-effort has to be accelerated. Our group is working on the mathematical modeling on cancer progression including different stages with applications to various types of cancers. Albeit cancer differs even between patients, some underlying mechanisms are comparable and therefore we generalize the similarities and simulate the biophysical processes by using analogous mathematical frameworks and numerical methods that are presented in the following section. Subsequently, several models and results are shown and conclusions, as well as prospects, are given.

2 Mathematical Concepts

Mathematical modeling can reshape our view of cancer [44]. An in-depth understanding of any phenomenon associated with cancer is ultimately based on the biological interaction between different time–space scales. Simplified mathematical models and formalisms to describe real biological processes and phenomena can be developed to provide hypotheses and predictions, as well as quantify their impacts for experimentalists to focus on [68].

Generally, mathematical models are based on abstract relations or a set of equations that ensue from a description of the various processes involved. The solution of the equations can be used to describe the evolution of a biological system over time or to simulate properties near an equilibrium point. Some hypotheses are necessary during the process of mathematical modeling to define the model and to make the problem well-tractable. In our group, we encode the mathematical model under the premise of the following assumptions: (1) in most instances, all cells have a circular projection on a two-dimensional substrate or a spherical shape inside a three-dimensional ECM; (2) each cell has a viable or dead discrete state; (3) to reduce the CPU time, we consider a small number of cells in either an 2D or 3D domain of computation as representative for a larger cell system; (4) cells are treated as point sources for chemokines and for the strain energy density. Based on the above assumptions, we sketch the common basic principles of the mathematical models for the various stages of cancer development in the following bullet points:

- Typically, cellular processes like **cell proliferation, cell death** and **cell mutation** happen thought to be caused by the interaction of internal genes and the external environment. Due to some unpredictable factors, we incorporate stochastic processes in the model as one of the criteria to determine whether division, death or mutation happens or not. The principle we are using is under the assumption that the probability of each process to occur is taken from a memoryless distribution and the probability can be influenced by mechanical force, chemokine concentration or other signals, which is given by Vermolen et al. [123]. With the probability per unit of time (that is probability rate) λ over a time interval Δt, then

$$
\begin{aligned}
P(t \in (t_n, t_n + \Delta t)) &= \int_{t_n}^{t_n + \Delta t} \lambda e^{-\lambda(t - t_n)} dt \\
&= 1 - e^{-\lambda \Delta t} \\
&\approx \lambda \Delta t.
\end{aligned} \tag{1}
$$

- **Tensotaxis** means that cells move according to mechanical tensions. The cells both experience a strain energy density field, and exert forces on their neighboring environment, which gives a contribution to the entire strain energy density field and which is essential for cell adhesion to effectuate cell migration, shape maintenance as well as other intercellular communications [92, 125]. Then the deformation of the substrate or ECM results into a strain energy which is transmitted to neighboring cells [72]. Moreover, the cells are not allowed to overlap too much because of the cell contact inhibition and therefore a repulsive force is taken into account. Having n cells exerting pulling forces and pushing forces if they are in physical contact, this gives at cell i

$$
M(\mathbf{x}_i) = \sum_{j=1}^{n} M_j^0 exp\{-\lambda_j \frac{\| \mathbf{x}_i - \mathbf{x}_j \|}{R}\} - \frac{4}{15\sqrt{2}} \frac{E}{\pi} \left(\frac{h}{R}\right)^{\frac{5}{2}}, \quad for\, i, j \in \{1, \ldots, n\}, \tag{2}
$$

where M and E represent the strain energy density and the Young's modulus. Further R, h, and λ_j, respectively, denote the cell radius, indentation distance, and mechanical signal attenuation ratio of the strain energy density through the substrate. Furthermore, $\mathbf{x}_i$ denotes the spatial location of the midpoint of cell i. The first term in the above equation represents the long-distance attenuation of the strain energy density over the extracellular matrix. The second term takes into account the contact forces exerted by the cells when they mechanically impinge. The second term in the right-hand side of the above equation is only nonzero if $||\mathbf{x}_i - \mathbf{x}_j|| < 2R$. More information regarding this formalism can be found in [22, 119].

- **Cell migration driven by chemical cues** includes chemotaxis and haptotaxis where cells move toward the concentration gradient of chemokines in a fluid phase or an extracellular matrix, respectively. Reactive T-cells move and fight against cancer cells directed by the gradient of tumor-derived chemokines [61, 114]. In the solid-part of ECM, the balance of chemokines typically looks like

$$\frac{\partial c}{\partial t} - D_c \Delta c = \sum_{j \in \mathbb{T}(t)} \gamma_j(t) \delta(\mathbf{x} - \mathbf{x}_j(t)), \quad j \in \mathbb{T}(t), \tag{3}$$

  where c, D_c, $\mathbf{x}$, and γ_j denote concentration and diffusivity of chemokine as well as location and secretion rate by cancer cells, respectively. Moreover, $\delta(\mathbf{r})$ represents the Dirac Delta Function to describe the secretion of chemokines for each cancer cell and $\mathbb{T}(t)$ represents a set of cancer cells which secrete chemokines. Further, $\mathbf{x}_j(t)$ denotes the center position of cell j at time t. Some more complicated chains of reactions like in [10] are possible.

- Over a broad type of cells, the cell boundary is mobile accompanied by a locally persistent random character [67] caused by **cellular random processes** and by random local properties of the ECM in terms of orientation and denseness. Thereby, we use a Wiener process $d\mathbf{W(t)}$ (also known as Brownian motion, where $d\mathbf{W}$ contains independent samples from the normal distribution $N(0, dt)$) to simulate the randomness in cell migration. Therefore, one cell migrates within the domain of computation by random walk, chemotaxis, durotaxis and as a result of the localized strains they are exposed to. Let $\mathbf{x}_j(t)$ denote the center position of cell i at time t, then this gives

$$d\mathbf{x}_i(t) = \alpha_i \hat{M}_i(\mathbf{x})\hat{\mathbf{z}}_i dt + \beta \nabla c(t, \mathbf{x}_i(t))dt + \sqrt{2D}d\mathbf{W(t)}, \quad for \ i \in \{1, \ldots, n\}. \tag{4}$$

  Where α, β, and D denote a parameter with dimension $[\frac{m^3}{Ns}]$, cell mobility, and cell diffusion coefficient. The first, second, and third term, respectively, represent tensotaxis, chemo- or haptotaxis, and random walk. More information regarding this formalism can be found in [22].

- Due to the **uncertainties from stochastic processes** in the model and **input parameters**, statistical assessment of the model and variables is very important. Monte Carlo simulations are performed to investigate the models and to estimate the influence of data, where the sensitivity of a couple of variables are tested simultaneously and quantitatively. To start with, random input values based on their probability distributions that can be lognormal, uniform, normal or other sorts of distributions need to be generated. Taking a normal distribution as an example, a stochastic variable X can be generated by Eq. 5 (see below), where N, σ, and μ denote the number of samples, the standard deviation, and the mean of the distribution, respectively. Subsequently, one repeats the calculations with N samples, and then constructs a frequency distribution [71, 79]. The results of Monte Carlo simulations become more accurate as the amount of trials increases, with an error that decreases inversely proportional to the square root of the number of samples. Moreover, the correlation coefficient r between several variables can be computed to establish whether statistically significant correlations between the variables of interest exist. This has been done for the transmigration of cancer cells through cavities and arteries in [21].

$$X = (randn(N, 1) \times \sigma) + \mu \tag{5}$$

- During cell migration, certain cells (e.g., cancer cells, immune cells) exhibit **ameoboid migration** where they push the surrounding ECM away and squeeze through constrictions, which are much smaller than their own diameters [130]. Furthermore, cells could secrete proteinases to degrade the ECM and remodel the surroundings [16]. In both scenarios, cells migrate through confined spaces accompanied by a morphological evolution. Since the cells deform when migrating through confined spaces, we take the evolution of the cell shape into account. In order to so, the cell boundary is divided into mesh points, where for each mesh point i an equation of motion is formulated on the basis of taxis, random motion and cell stiffness. This equation reads as

$$d\mathbf{x}_i(t) = \beta \nabla c(t, \mathbf{x}_i(t))dt + \alpha(\mathbf{x}_i^n(t) + B(\phi)\hat{\mathbf{x}}_i - \mathbf{x}_i(t))dt + \eta d\mathbf{W(t)}, \quad for\ i \in \{1, \ldots, n\}, \tag{6}$$

where β represents the cell's response to external signals and $\alpha > 0$ stands for the cytoplasm's stiffness. Since the cell boundary is divided into points that are connected to the nucleus, the $\mathbf{x}_i(t)$ and $\mathbf{x}_i^n(t)$, respectively, denote the nodal points on the cell membrane and nucleus surface. Here, $\hat{\mathbf{x}}_i$ is a vector connecting the initial positions of one point i on the cell membrane to the one point i on nucleus surface. Furthermore, $d\mathbf{W(t)}$ denotes a vector Wiener process multiplied to model local random mobility. Since we incorporate the deformation of the cells, it is important to simulate the correct angular orientation of the cell, we incorporate the rotation matrix $B(\phi)$,

$$B(\phi) = \begin{pmatrix} \cos(\phi) & -\sin(\phi) \\ \sin(\phi) & \cos(\phi) \end{pmatrix}, \tag{7}$$

where ϕ is the angle of rotation determined to ensure the following position of one cell is as close as possible to the current position during the adjustment of cell orientation. More details regarding the cell (and nucleus) deformation model and the three-dimensional version of the rotation matrix as well as applications are given in [21].
- With intravasation and extravasation of blood vessels or lymphatic vessels, individual cancer cells or multiple cells are able to transmigrate from part to part and thereby seed into distant organs. One way of transmigrating is through blood ves-

sels. First, the cancer cells penetrate through a vessel wall to get into the blood vessel. Subsequently, they are advected and transmigrate out of the blood vessel through the vessel wall. Having arrived in the new body part, the cancer cell is sometimes able to set up a new colony. In this case, a steady **blood fluid flow** is necessary to be incorporated. To model a laminar flow as well as to simplify the process, the pressure-induced Poisseuille flow serves as a method, which reads as

$$u_z(r) = -\frac{\partial p}{\partial z} \cdot \frac{R_t^2}{4\mu} \cdot (1 - \frac{r^2}{R_t^2}), \quad in \; \Omega_b, \tag{8}$$

where R_t, μ and p represent half-width of the blood vessel, viscosity, and pressure. One cell with a radius of r moves with a parabolic profile velocity in the vasculature domain Ω_b. Some computational models can be found in [23, 122].

2.1 Applications

We have already said earlier the scope of our models involves various stages of cancer progression consisting of tumor initiation, immune responses, angiogenesis, and metastasis. Since the current paper aims at describing the general applicability of cell-based mathematical models to simulate cancer progression, we show and discuss some of the simulation results from the various models. More specific details can be found in the respective papers.

2.2 Modeling Tumor Initiation

Vermolen et al. [123] develop a semi-stochastic cell-based model to simulate tumor initiation in two and three dimensions. In the model, the mutated cancer cells are allowed to spread and to invade the neighboring tissue, which induces chemotaxis-driven migration of antitumor T-lymphocytes. Then T-cells transmigrate from adjacent blood vessels and move in the direction of to the concentration gradient of chemokines and cytokines secreted by cancer cells. This simulation takes into account the proliferation and death by apoptosis of epithelial and cancer cells, mutation of epithelial cells and immune responses of antitumor T-lymphocytes. In the simulations, the computational domain is modeled as spherical where on designated locations T-cells appear randomly as a result of excavation from surrounding blood vessels.

In Fig. 2, we show several snapshots at consecutive times from a simulation of the initiation of cancer. The simulations are done in an 3D framework. The spherical domain with a radius of 40 micrometer is filled with endothelial cells (indicated in green color), which are allowed to migrate, proliferate as well as being subject to apoptosis, at the beginning. Once the endothelial cells mutate to cancer cells

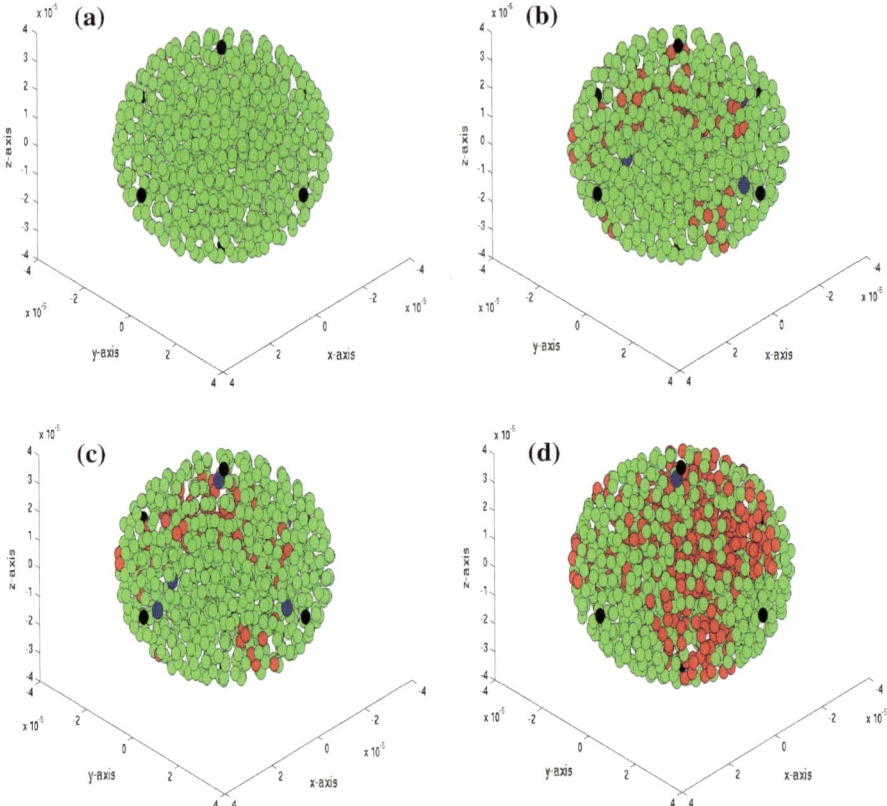

Fig. 2 Tumor development at four consecutive times from a front view. The green, red, and blue cells represent, respectively, the epithelial cells, tumor cells, and immune T-cells. The black dots represent locations on the blood vessel points where T-cells are able to extravasate freely [123]

(indicated in red color), antitumor T-cells (indicated in blue color) are able to be released and to migrate from the blood vessels, which are depicted by six black dots. The number of cancerous cells increases significantly in the domain, whereas the T-cells, represented by blue spheres, hardly enter the region from the boundary. The intercellular contact inhibition force inhibits the chemotaxis migration of T-cells toward the center. Therefore, under this simulation, one can imagine that the number of cancerous cells gradually increases and that the cancerous cells will dominate in the tissue or have a detrimental impact on an organ of a patient if the immune system is not sufficiently strong in terms of T-cells counts and migration rate (mobility).

This model captures the most relevant aspects of the tumor initiation which lays the foundation for the further understanding of the microscopic phenomenon as well as for the exploration of drug treatment to fight the cancer cells optimally. We finally remark that the computational limitations forced us to consider a very small portion of the tissue with a size order of several tens of micrometers. Further, the simulation can be used to simulate the very early stages of general cancers such as lung cancer where cancer cells have easy access to abundant levels of oxygen.

2.3 Modeling of T-cells Migration in Pancreatic Cancer

The tumor-specific T-lymphocytes play an essential role in antitumor immune responses, which is the body's first line of defense to clear the body from pathogens, hazardous chemicals, and cancerous cells. However, cancer cells could escape their removal by T-cells via several ways. For example, pancreatic ductal adenocarcinoma is notorious for its profuse desmoplastic ECM, which is a barrier built by cancer cells to protect themselves against T-cells [97]. This anisotropic stromal ECM is arranged parallel to the tumor circumference, hence making it appear like tumor islets [100]. The cross-talk between a solid tumor and its microenvironment as well as the function of anisotropic stromal ECM are unclear due to the dynamic changes in cellular and noncellular constituents [6, 25, 49, 84]. However, this anisotropic stroma does have an obstructive effect on T-cells migration in antitumor immune responses, where T-cells mainly move between two parallel collagen fibers and hence they migrate faster in the direction of the orientation of fibers, once they enter the stromal ECM [12]. Thence, the reduction in their entry reduces their ability to engulf and remove the cancer cells.

The influence of anisotropic ECM networks on T-cells migration in and around pancreatic tumor islets has recently been studied by Chen et al. [22]. In this aforementioned work, we are able to quantify the delay of T-cells invasion in cancer-affected area due to anisotropic ECM orientation. Since anisotropic ECM has an effect on limiting the direction and the magnitude of the velocity component of T-cells in the direction toward the center of the tumor islets around the pancreatic tumor islets, T-cells are hindered in their entering the tumor islets. In Fig. 3, a thick annular gray domain depicts an anisotropic ECM filled with profuse collagen as well as other noncellular constituents. Inside the tumor islets, the epithelial cells (indicated in blue) are able to migrate, proliferate, die or mutate to cancer cells (the latter phenotype is indicated in red).

With an increase of chemokine signals secreted by cancer cells, T-cells (indicated in green) migrate in the direction of the inner side to fight against cancer cells. As a result, T-cells engulf all the cancer cells in the tumor islets under a strong immune system (see Fig. 3a, c, where a "strong immune system" refers to the case that the immune system is sufficiently strong to neutralize the cancer cells), whereas T-cells are not able to control the number of cancer cells under a weak immune system (see Fig. 3b, d). Furthermore, the model quantifies the increase in the length of the period for the T-cells needed to engulf the cancer cells under both isotropic and anisotropic circumstances (see Fig. 4). In terms of the obstructing effect of anisotropic ECM on T-cells migration, the higher the degree of anisotropy, the harder it becomes for the T-cells to enter inside.

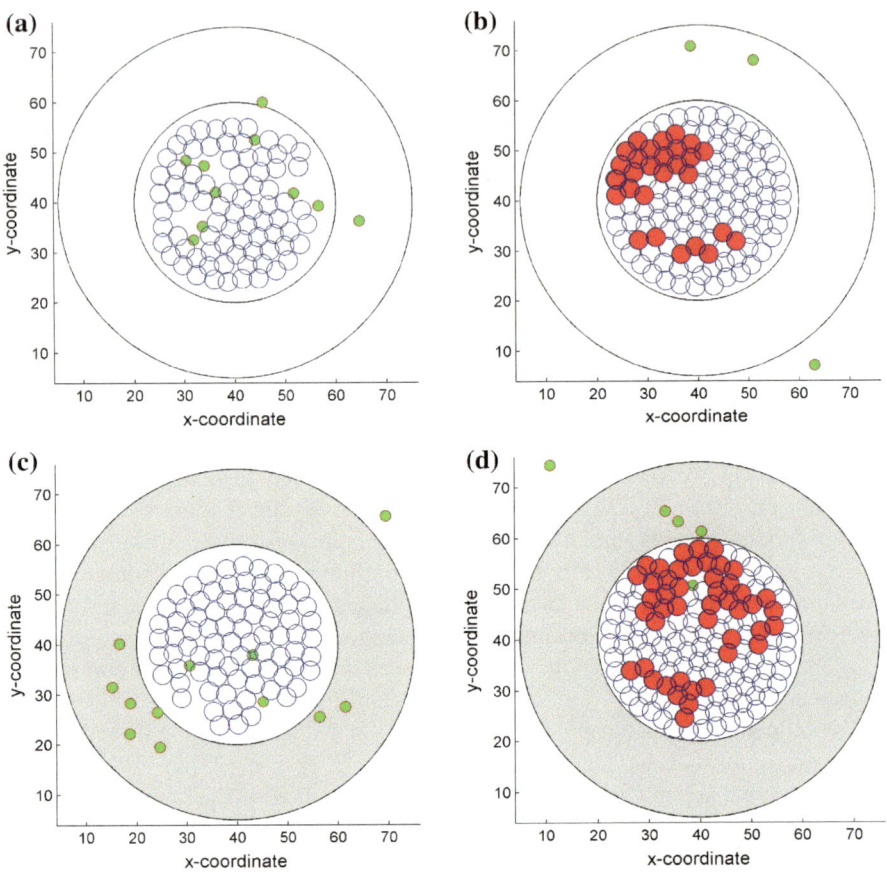

Fig. 3 Tumor islets **a**, **b** with an isotropic and with **c**, **d** an anisotropic ECM orientation under a weak **b**, **d** and a strong **a**, **c** immune reaction. The blue, red, and green cells represent epithelial cells, cancer cells, and T-cells, respectively [22]

One important objective of this model is to set up a formulism for the inhibition of adaptive immune system caused by the orientation of anisotropic ECM layer. It would be meaningful to combine the model with experiments to investigate lethal pancreatic cancer in the early stages. Moreover, the model could also be used to quantify the exploration of drug treatment and to quantify the needed strength of the immune system in terms of T-cells counts, T-cells migration speed, and engulfment capacity for a certain aggression (in terms of cancer cell proliferation rate and mutation rate) of cancer proliferation.

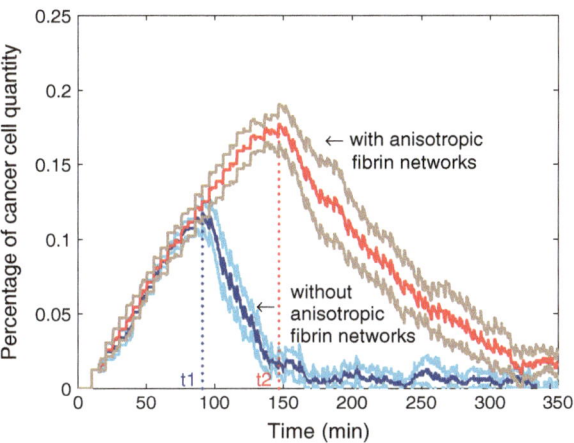

Fig. 4 Comparison of the evaluation of the percentage of cancer cell amount. The blue line and red line denote the tumor islets under a strong immune system without (with immune time t_1) and with (with the immune time t_2) anisotropic ECM orientation. The light blue and brown lines represent the corresponding 95% confidence intervals [22]

2.4 Modeling Angiogenesis

Angiogenesis is the process of new blood vessels sprouting from existing ones. This process includes degradation of vascular basement membrane, activation, proliferation, and migration of endothelial cells, as well as reconstruction of vascular networks. This is a normal process in the body that is also employed by growing tumors as their blood supply demands grow. The work by Bookholt et al. [10] takes into account the dynamic interaction during angiogenesis between two kinds of endothelial cells: stalk cells and tip cells.

Bookholt et al. [10] simulate early angiogenesis in a scaffold by using a cell-based formalism combined with a finite element method in three dimensions to solve the partial differential equations for the chemical entities. Their work could be fruitfully applied to model angiogenesis near a tumor. The migration of endothelial cells (ECs) proceeds via chemotaxis and durotaxis which is modeled by obtaining the numerical solution of a system of stochastic differential equations in the form of Eq. 4 and a set of diffusion-reaction equations of the form of Eq. 3. This model is expected to be applicable to simulate a re-establishment of a vascular network around cardiac arteries or other organs, as well as inhibiting the regeneration of tumor blood vessels.

The resulting mathematical problem is solved by a combination of a cell-based approach with the finite element method, the results are shown in Fig. 5. In Fig. 5a, the green and red cells denote the stalk and tip cells, respectively. The tip cells take the lead and move faster toward the bottom which can be observed from different angles. The tip cells rush to the front to chemically degrade the basement membrane and fibrin by which channels are formed. Endothelial cells are able to degrade the collagen and fibrin matrix by releasing MMPs as well as uPA [59]. The interplay between cancer cells, uPA, uPA inhibitors, plasmin, and ECM is numerically simulated by Andasari et al. [4]. Figure 6a shows the first layer at the top surface of the experiment, which had to be considered since the experimental study, see Fig. 6b, measured the area of

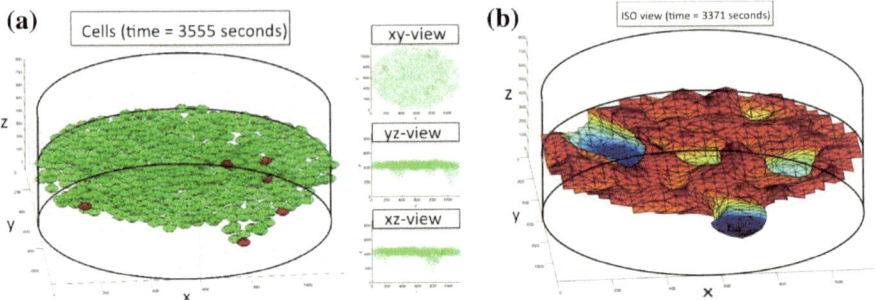

Fig. 5 **a** Three-dimensional cell plot on the left and projections with respect to different angles of plot in right. The green and red cells denote the stalk and tip cells, respectively. **b** Three-dimensional surface plot of cells. The cylindrical boundary in black lines denotes the computational domain [10]

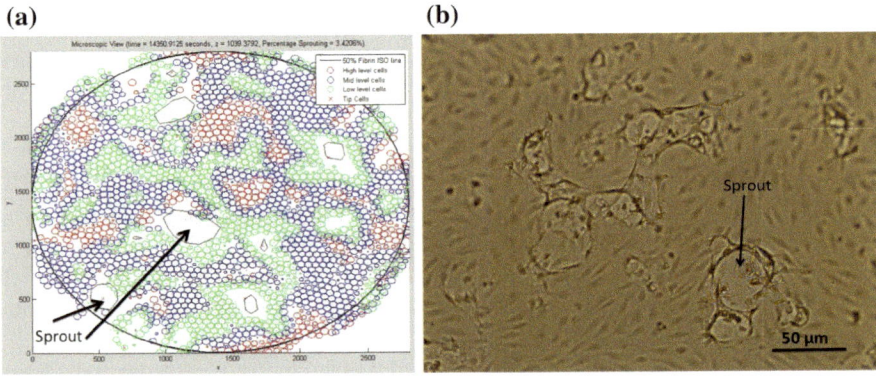

Fig. 6 **a** Microscopic plot from a top view. The red, blue, and green circles represent cells in top, middle, and lower levels. Furthermore, the small red crosses are tip cells and white gaps marked with black arrows visualize new vessel sprouts. **b** Dermal ECs in a well after stimulation with 25 ng/mL vascular endothelial growth factor (VEGF) and 2 ng/mL the cytokine named tumor necrosis factor α (TNF-α). Similar to the circular shape of the structures are new vessel sprouts, and one of which is marked by an arrow [10]

the gaps of the top surface. Figure 6b represents a micrograph showing the top layer of the experimental gel structure through which the endothelial cells migrate. The gaps represent the channels that resulted by endothelial cells migrating into the gel and their simulated counterparts can be observed in Fig. 6a. The channels, formed by the tip cells-induced degradation of the ECM, are represented by the white patches over the surface in Fig. 6a, and by the areas enclosed by the closed curves in Fig. 6b. Furthermore, Fig. 6a shows the other cells in the very top layers of top, middle and lower plotted by red, blue as well as green, respectively. Two gaps that coincide with vessel sprouts formed by the tip cells, are indicated by arrows in Fig. 6a. This cell-based model is qualitatively successful in describing the in vitro angiogenesis sprouting experiments done by the VUmc dermatology department (see Fig. 6b).

2.5 Modeling of Durotaxis-Driven Migration of Cancer Cells in Metastasis

Cancer cells are able to apply relatively large forces [65, 72] which relate to their increased mobility and mechanical invasiveness [3, 76]. The study by Dudaie et al. [32] is devoted to the development of a model for collective migration of cells under varying stiffness and cell-to-cell mechanical communication. The simulation results indicate that cells prefer to move to a stiffer region and that the rate of migration depends on the stiffness of the cell and substrate or ECM. This model is developed to simulate cancer metastasis, in particular in the context of breast cancer, and other processes such as wound healing. In the aforementioned work, the stiffness of the cells and substrate or ECM is assumed to be fixed at all times in an 2D or 3D model. However, the tumors are stiffer due to a stiff stroma compared to normal tissue [85]. Furthermore, cancer cells are capable of changing the stiffness of their environment. This is the reason why the influence of varying stiffness on cell migration is investigated by Dudaie et al. [32]. Based on their results, the migration velocities of the cells differ with varying elastic modulus of the substrate and the simulation can be used for describing cancer metastasis. See Fig. 7 for several snapshots at consecutive times. Furthermore, the ratio between the elastic moduli of the substrate and the cell steers the net direction of cell migration. In Fig. 7, the blue, red, and black cells represent the normal cells, cancerous cells, and the dead cells, respectively. Due to the attraction by the tumor, normal cells in the tissue tend to move toward the tumor domain where the ECM is much stiffer than it is in the normal tissue. Cells have previously been shown to be attracted to stiffer substrates [66].

This model focuses on the assessment of the influence of the stiffness on cellular migration through mechanotaxis and as a result, cells are likely to move toward a stiffer region (that is durotaxis). Besides metastasis, this model can also be applied to other biomechanical processes, such as tissue growth, wound healing, tumor development, or preventing migration from metastasis.

2.6 Modeling Cell Deformation During Cancer Metastasis

Next to cell colony models with relatively many cells, we also consider the behavior of one cell only. This is done experimentally [65, 72] and computationally. This single cell approach allows to study the behavior of one cell in more detail and this could help to reveal the migration and transmigration kinetics through cavities and other objects.

In most of our models, the shape of a cell is fixed to circular in 2D and spherical in 3D; this is the morphology of cells in suspension and also on soft gels. However, the real shape of a cell and nucleus during migration could be dynamic [42]. To simulate dynamic cell shape changes, a phenomenological model with application to simplified cancer metastasis has been developed [21]. One cell migrates toward to

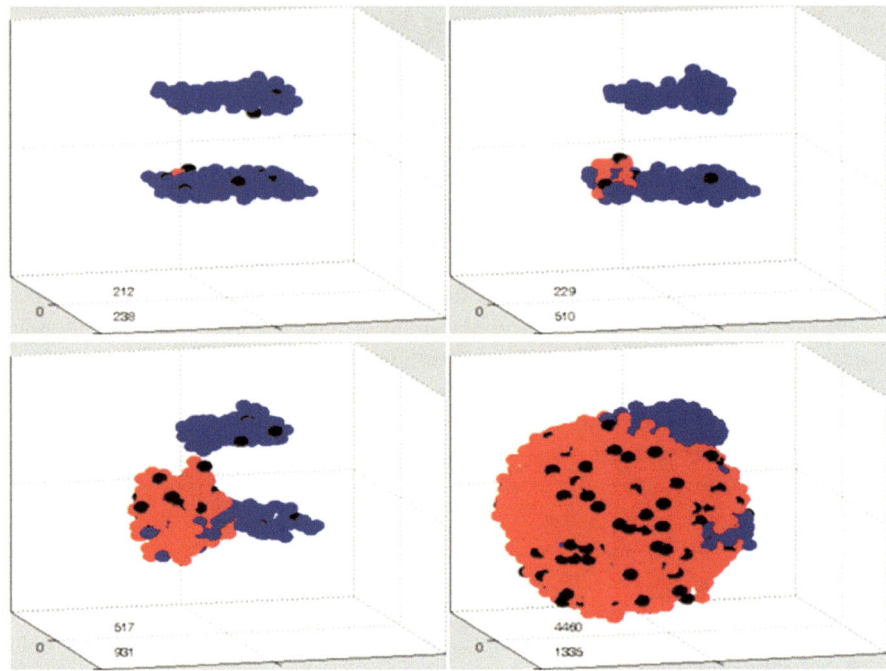

Fig. 7 3D simulation of metastasis of cancer. The blue, red, and black cells represent the normal cells, cancer cells, and dead cells, respectively [32]

an imaginary source in Fig. 8, which could be the oxygen emitting spot or a stiffness signal. We use this model to simulate how a single cell intravasates and extravasates a blood vessel or a lymphatic vessel subject to the evolution of morphology caused by constricted spaces. To mimic a micro blood fluid, Poisseuille flow is incorporated to simulate blood flow through a small blood vessel in an extended model where the impact of vessel radius on the success rate of cell transmigration is studied [23].

We use an IMEX Euler Maruyama method to deal with the nonlinear stochastic differential equations and Monte Carlo simulations to analyze the correlations between uncertain input values and simulated results. In our simulations, a cell is able to deform well when it comes across various stiff obstacles in 2D or 3D surroundings. This is consistent with various in vitro experimental studies. Moreover, Monte Carlo simulations show that, among others, the correlation between the first passage time, which is the time needed for the cell to reach the hypothetical source (denoted by the asterisk), and the channel roughness is significantly positive.

This model could be used to find ways to inhibit metastasis through inhibition of transmigration of cancer cells through blood vessels and organ walls. One could for instance search for ways to reduce the metastatic driving force by decreasing the signal for durotaxis by reducing the environmental stiffness variations, or by exposing the cancer cells to a chemical that decrease their strength in applying forces on their

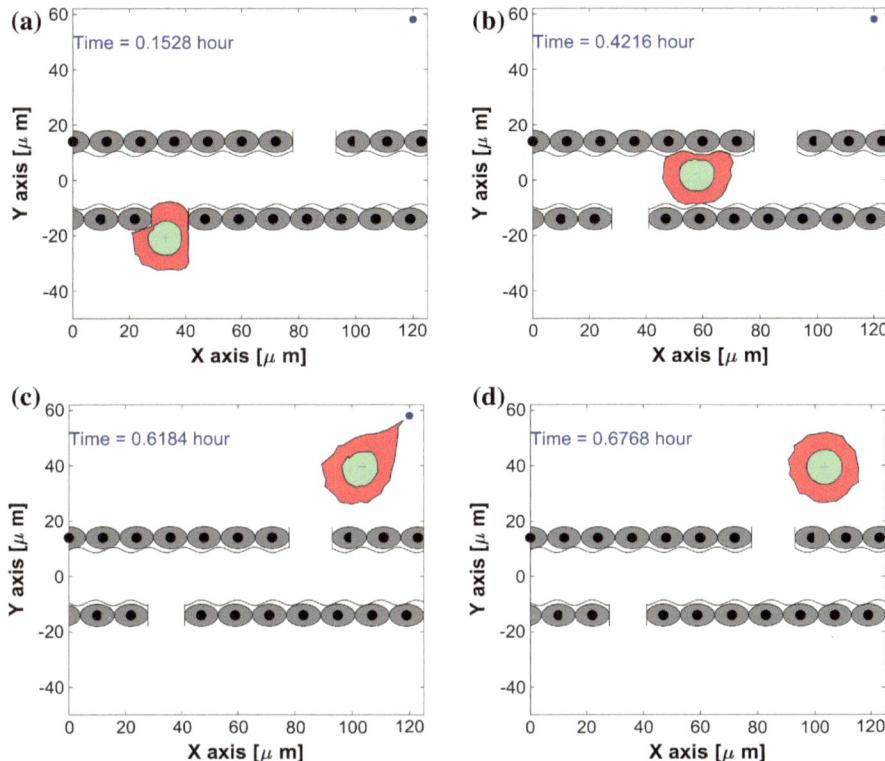

Fig. 8 Consecutive snapshots of one cell deformation during intravasation and extravasation. The vessels, metastatic cell, and nucleus are visualized in gray, red, and green colors, respectively. The cell is attracted by a hypothetical source (e.g., a chemical signal) represented by a blue asterisk [21]

extracellular environment or by reducing the deformability (and motility) of the cells such that the cancer cells are no longer able to move from one part in the body to another part through cavities and other types of obstacles. Since the mathematical modeling of biomedical processes only started relatively recently, many challenges remain to be dealt with.

3   Discussion

Mathematical simulations with a wide range of techniques can be applied to many different practical problems and its importance on cancer research has been increasingly recognized in the recent decades. Differential equation-based continuum models are able to cover relevant scales from 10^2 μm to 10 cm, whereas hybrid models including cellular automata, and agent-based techniques can span scales from microns to millimeters [103]. Normally, mathematical modeling is a process of several steps: (1) choose a specific problem and computational domain; (2) make some simplifications using assumptions and convert this real problem to a mathematical problem through quantification; (3) establish mathematical equations which enable to describe the relationship between the relevant quantities; (4) calculate the solution to the actual problem quickly and accurately using computing technology, software and other tools. Certainly, a sound mathematical model in this context must be analyzed (such as well-posedness, stability, error, *etc.*), validated *in vitro* or *in vivo* and applied to obtain further understanding of the fatal disease or any other biophysical or biomedical phenomenon.

Since biomedical studies often involve extensive experimental data in terms of patterns and numbers, the quantified hypotheses pose mathematical challenges, which are the backbone of mathematical models. The mathematical models can be used to investigate case studies that do not exactly fit within the experimental outcomes. A major advantage of mathematical modeling is that the number of animal or in vitro experiments can be reduced. Many mathematical models are based on the abstraction of biological phenomena into sets of partial differential equations, stochastic processes or even combinations of both. The approximate solutions are solved by using numerical methods such as combinations of time integration, finite element method, Green's functions *etc.*, or using stochastic processes. In case of probabilistic models, the model results need a statistical assessment in terms of intervals of confidence, correlations, or other statistical tests.

In our future studies, we will develop models with expanded physiological, such as the formation of abnormal stroma caused by cancer, interaction of innate immunity and adaptive immunity for cancer, angiogenesis, and network models for cancer metastasis, as well as other processes. Through the combination of various stages of the models of cancer, the complete model is expected to be applied in various aspects of cancer research. To keep the CPU time of the model low, a small number of cells is considered in two and three dimensions currently. However, parallel computing is capable to upscale the number of cells in size to achieve large-scale quantitative simulation of cells. Therefore, we will use parallel computing facilities to simulate human tissue and even organs in an 3D environment to make the model as realistic as possible.

Existing models by us and others will also need to be expanded to include more complexity and physiological aspects. One aspect is sensitivity analysis of input values, due to the uncertainties, the study of parametric variation is crucially important. In our statistical evaluation of results, we use Monte Carlo simulations, which enables us to simultaneously and quantitatively investigate the input variables and correlations among them [71, 79]. It could be useful to compare our statistical outcomes in terms of probabilities, correlations and significance with large data sets from experimental and clinical studies and to investigate whether similar trends arise.

Another important matter is the accessibility to realistic values of the input parameters, currently, most of the parameters in our simulations have been chosen on the basis of the literature or by educated guesses. Hopefully, more realistic input values will be available to us such that we are able to validate our modeling results better with available experimental outcomes. Thence it is crucial to cooperate with biomedical labs or hospitals to realize the validation, evaluation, and application of our models, which definitely enable to enhance further understanding of the progression and inhibition of cancer. Another important issue concerns the variations of the input from patient to patient due to age, genetic, pattern, lifestyle, and gender. This makes that many of the simulated results contain uncertainties despite possible well-measured data in generic patients. These uncertainties require a probabilistic modeling approach and hence, a statistical assessment of the simulation results is indispensable.

Last but not the least, next to common surgical therapies where tumors are removed or where chemotherapy is applied, therapies could be directed to paralyzing cancer cells in terms of motility and invasiveness by reducing cell deformability and/or by reducing the durotactic signal through de-stiffening certain body parts. Further treatments could target cancer cells by decreasing their proliferation rate and by increasing mortality rates. This is often done in chemotherapies. Alternatively, one could investigate and quantify the treatment of cancers by nanoparticles that only target the cancer cells. Modeling studies and frameworks could help investigate the impact and feasibility of the aforementioned treatments.

4 Conclusion

In the current manuscript, we illustrated the importance of mathematical modeling to the cancer research community. The illustration is done by describing several case studies of models of various applications in early-stage cancer development. A large advantage of mathematical modeling is the availability of a tool to predict outcomes from conditions that are beyond the measured and observed values. Biological and clinical researchers normally have limited training in mathematics, conversely, applied mathematicians often poorly understand the complicated multiscale dynamics that characterize the processes studies in the life sciences [45]. As a fact, typically medical biologists and clinicians show little interest in mathematical modeling, and thereby limited data is accessible to mathematicians. Another problem is that often mathematical modelers ask for parameters that are hard or even impossible to measure by medical biologists.

The evolutionary nature of cancer is undoubtedly important for oncologists to hack cancer disease, where mathematical modeling is able to aid us to obtain a better understanding of how cancer evolves and how it adapts to the environment. Moreover, mathematical modeling can help clinicians optimize the drug treatment strategies, and further do the pre-validation studies on a computer with a few seconds before testing in animals or humans [44]. At later stages, the mathematical models will be used to improve existing therapies and to quantify the impact of new therapies against cancer. To summarize, mathematical modeling has brought new insights into the underlying mechanisms of cancer evolution and provides prospects for oncology research.

Acknowledgements This study is financially supported by the China Scholarship Council and the authors are very grateful for this funding. The authors declare that they do not have any conflicts of interest.

References

1. Abbott LH, Michor F (2006) Mathematical models of targeted cancer therapy. Br J Cancer 95(9):1136–1141
2. Ahmadzadeh H, Webster MR, Behera R, Jimenez AM, Valencia DW, Weeraratna AT, Shenoy VB (2017) Modeling the two-way feedback between contractility and matrix realignment reveals a nonlinear mode of cancer cell invasion. Proc Natl Acad Sci 114(9):E1617–E1626
3. Alvarez-Elizondo MB, Weihs D (2017) Cell-gel mechanical interactions as an approach to rapidly and quantitatively reveal invasive subpopulations of metastatic cancer cells. Tissue Eng Part C Methods 23(3):180–187
4. Andasari V, Gerisch A, Lolas G, South AP, Chaplain MAJ (2011) Mathematical modeling of cancer cell invasion of tissue: biological insight from mathematical analysis and computational simulation. J Math Biol 63(1):141–171
5. Anderson ARA, Chaplain MAJ, Luke Newman E, Steele RJC, Thompson AM (2000) Mathematical modelling of tumour invasion and metastasis. Comput Math Methods Med 2(2):129–154
6. Angeli F, Koumakis G, Chen M-C, Kumar S, Delinassios JG (2009) Role of stromal fibroblasts in cancer: promoting or impeding? Tumor Biol 30(3):109–120
7. Nicola B, Luidgi P (2000) Modelling and mathematical problems related to tumor evolution and its interaction with the immune system. Math Comput Model 32(3–4):413–452
8. Fortunato B, Elisa B, Vienna L, Lucio C, Antonella F, Paolo V (2012) Computational model of egfr and igf1r pathways in lung cancer: a systems biology approach for translational oncology. Biotechnol Adv 30(1):142–153
9. Block M, Schöll E, Drasdo D (2007) Classifying the expansion kinetics and critical surface dynamics of growing cell populations. Phys Rev Lett 99(24):248101
10. Bookholt FD, Monsuur HN, Gibbs S, Vermolen FJ (2016) Mathematical modelling of angiogenesis using continuous cell-based models. Biomech Model Mechanobiol 15(6):1577–1600
11. Borau C, Polacheck WJ, Kamm RD, García-Aznar JM (2014) Probabilistic voxel-fe model for single cell motility in 3d. Silico Cell Tissue Sci 1(1):1–17
12. Bougherara H, Mansuet-Lupo A, Alifano M, Ngô C, Damotte D, Le Frère-Belda M-A, Donnadieu E, Peranzoni E (2015) Real-time imaging of resident t cells in human lung and ovarian carcinomas reveals how different tumor microenvironments control t lymphocyte migration. Front Immunol 6:500
13. Helen B, Dirk D (2009a) Individual-based and continuum models of growing cell populations: a comparison. J Math Biol 58(4):657–687
14. Helen B, Dirk D (2009b) Individual-based and continuum models of growing cell populations: a comparison. J Math Biol 58(4–5):657–687
15. Chaffer CL, Weinberg RA (2011) A perspective on cancer cell metastasis. Science 331(6024):1559–1564
16. Chieh C, Zena W (2001) The many faces of metalloproteases: cell growth, invasion, angiogenesis and metastasis. Trends Cell Biol 11:S37–S43
17. Chaplain Mark AJ (2000) Mathematical modelling of angiogenesis. J. Neuro-Oncol 50(1):37–51
18. Chaplain MAJ, McDougall SR, Anderson ARA (2006) Mathematical modeling of tumor-induced angiogenesis. Annu Rev Biomed Eng 8:233–257
19. Chuanying Chen B, Pettitt M (2011) The binding process of a nonspecific enzyme with dna. Biophys J 101(5):1139–1147

20. Chen J, Vermolen FJ (2016) Literature study on cell-based semi-stochastic modelling for the dynamics of growth of cell colonies
21. Chen J, Weihs D, Van Dijk M, Vermolen FJ (2018a) A phenomenological model for cell and nucleus deformation during cancer metastasis. Biomech Model Mechanobiol 17(5):1429–1450
22. Chen J, Weihs D, Vermolen FJ (2018b) A model for cell migration in non-isotropic fibrin networks with an application to pancreatic tumor islets. Biomech Model Mechanobiol 17(2):367–386
23. Chen J, Weihs D, Vermolen FJ (2018c) Monte carlo uncertainty quantification in modelling cell deformation during cancer metastasis. In: Proceedings of the CMBBE2018
24. Chmielecki J, Foo J, Oxnard GR, Hutchinson K, Ohashi K, Somwar R, Wang L, Amato KR, Arcila M, Sos ML et al. (2011) Optimization of dosing for egfr-mutant non–small cell lung cancer with evolutionary cancer modeling. Sci Transl Med 3(90):90ra59–90ra59
25. Clarijs R, Ruiter DJ, de Waal RMW (2003) Pathophysiological implications of stroma pattern formation in uveal melanoma. J Cell Physiol 194(3):267–271
26. European Commission (2016) Animals used for scientific purposes. http://ec.europa.eu/environment/chemicals/lab_animals/3r/alternative_en.htm
27. Couzin-Frankel J (2013) Cancer immunotherapy. Science 342(6165):1432–1433
28. Da-Jun T, Tang F, Lee T, Sarda D, Krishnan A, Goryachev A (2004) Parallel computing platform for the agent-based modeling of multicellular biological systems. PDCAT. Springer, Berlin, pp 5–8
29. Micah D, Yu-Li W (1999) Stresses at the cell-to-substrate interface during locomotion of fibroblasts. Biophys J 76(4):2307–2316
30. Dirk D, Stefan H (2005) A single-cell-based model of tumor growth in vitro: monolayers and spheroids. Phys Biol 2(3):133
31. Dirk D, Stefan H (2005) A single-cell-based model of tumor growth in vitro: monolayers and spheroids. Phys Biol 2(3):133
32. Dudaie M, Weihs D, Vermolen FJ, Gefen A (2015) Modeling migration in cell colonies in two and three dimensional substrates with varying stiffnesses. Silico Cell Tissue Sci 2(1):2
33. DuFort CC, Paszek MJ, Weaver VM (2011) Balancing forces: architectural control of mechanotransduction. Nat Rev Mol Cell Biol 12(5):308–319
34. Elliott CM, Stinner B, Venkataraman C (2012) Modelling cell motility and chemotaxis with evolving surface finite elements. J R Soc Interface 9(76):3027–3044
35. Enderling H, Anderson ARA, Chaplain MAJ (2007a) A model of breast carcinogenesis and recurrence after radiotherapy. PAMM 7(1):1121701–1121702
36. Enderling H, Anderson ARA, Chaplain MAJ, Munro AJ, Vaidya JS (2006) Mathematical modelling of radiotherapy strategies for early breast cancer. J Theor Biol 241(1):158–171
37. Enderling H, Chaplain MAJ, Anderson ARA, Vaidya JS (2007b) A mathematical model of breast cancer development, local treatment and recurrence. J Theor Biol 246(2):245–259
38. Fearon ER, Vogelstein B (1990) A genetic model for colorectal tumorigenesis. Cell 61(5):759–767
39. Ferlay J, Soerjomataram I, Ervik M, Dikshit R, Eser S, Mathers C, Rebelo M, Parkin DM, Forman D, Bray F (2014) Globocan 2012 v1.0, cancer incidence and mortality worldwide: Iarc cancerbase no. 11 [internet]. 2013; Lyon, France: International agency for research on cancer. globocan.iarc.fr/Default.aspx
40. Folkman J, Haudenschild C (1980) Angiogenesis in vitro
41. Forkman J (1974) Tumor angiogenesis: Role in regulation of tumor growth. Syrup Soc Dev Biol 30:43–52
42. Peter F, Katarina W, Jan L (2011) Nuclear mechanics during cell migration. Curr Opin Cell Biol 23(1):55–64
43. Friedman R, Boye K, Flatmark K (2013) Molecular modelling and simulations in cancer research. Biochim Biophys Acta (BBA)-Rev Cancer 1836(1):1–14
44. Katharine G (2012) Mathematical modelling: forecasting cancer. Nature 491(7425):S66–S67

45. Gatenby RA (2010) Mathematical modeling in cancer. Biomedical informatics for cancer research. Springer, Berlin, pp 139–147
46. Gatenby RA, Vincent TL (2003) An evolutionary model of carcinogenesis. Cancer Res 63(19):6212–6220
47. Glazier JA, Graner F (1993) Simulation of the differential adhesion driven rearrangement of biological cells. Phys Rev E 47(3):2128
48. Hanahan D, Weinberg RA (2000) The hallmarks of cancer. Cell 100(1):57–70
49. Hanahan D, Weinberg RA (2011) Hallmarks of cancer: the next generation. Cell 144(5):646–674
50. Hatzikirou H, Breier G, Deutsch A (2014) Cellular automaton modeling of tumor invasion. Encyclopedia of complexity and systems science. Springer, Berlin, pp 1–13
51. Haralambos H, Andreas D (2008) Cellular automata as microscopic models of cell migration in heterogeneous environments. Curr Top Dev Biol 81:401–434
52. Haralampos H, Andreas D, Carlo S, Matthias S, Kristin S (2005) Mathematical modelling of glioblastoma tumour development: a review. Math Model Methods Appl Sci 15(11):1779–1794
53. Wonpil I, Stefan S, Benoit R (2000) A grand canonical monte carlo-brownian dynamics algorithm for simulating ion channels. Biophys J 79(2):788–801
54. Jackson TL (2004) A mathematical model of prostate tumor growth and androgen-independent relapse. Discret Contin Dyn Syst Ser B 4(1):187–202
55. Jemal A, Bray F, Center MM, Ferlay J, Ward E, Forman D (2011) Global cancer statistics. CA: Cancer J Clin, 61(2):69–90
56. Jemal A, Murray T, Ward E, Samuels A, Tiwari RC, Ghafoor A, Feuer EJ, Thun MJ (2005) Cancer statistics, 2005. CA: Cancer J Clin 55(1):10–30
57. Jeon J, Quaranta V, Cummings PT (2010) An off-lattice hybrid discrete-continuum model of tumor growth and invasion. Biophys J 98(1):37–47
58. Jolly MK, Boareto M, Debeb BG, Aceto N, Farach-Carson MC, Woodward WA, Levine H (2017) Inflammatory breast cancer: a model for investigating cluster-based dissemination. NPJ Breast Cancer 3(1):21
59. Kalebic T, Garbisa S, Glaser B, Liotta LA (1983) Basement membrane collagen: degradation by migrating endothelial cells. Science 221(4607):281–283
60. Kansal AR, Torquato S, Harsh GR, Chiocca EA, Deisboeck TS (2000) Simulated brain tumor growth dynamics using a three-dimensional cellular automaton. J Theor Biol 203(4):367–382
61. Kershaw MH, Wang G, Westwood JA, Pachynski RK, Lee Tiffany H, Marincola FM, Wang E, Young HA, Murphy PM, Hwu P (2002) Redirecting migration of t cells to chemokine secreted from tumors by genetic modification with cxcr2. Hum Gene Ther 13(16):1971–1980
62. Kim Y, Othmer HG (2013) A hybrid model of tumor-stromal interactions in breast cancer. Bull Math Biol 75(8):1304–1350
63. Kim Y, Stolarska MA, Othmer HG (2007) A hybrid model for tumor spheroid growth in vitro i: theoretical development and early results. Math Model Methods Appl Sci 17(supp01):1773–1798
64. Kirschner D, Panetta JC (1998) Modeling immunotherapy of the tumor-immune interaction. J Math Biol 37(3):235–252
65. Kristal-Muscal R, Dvir L, Weihs D (2013) Metastatic cancer cells tenaciously indent impenetrable, soft substrates. New J Phys 15(3):035022. https://doi.org/10.1088/1367-2630/15/3/035022/meta
66. Benoit L, Alice N (2012) Physically based principles of cell adhesion mechanosensitivity in tissues. Rep Prog Phys 75(11):116601
67. Lauffenburger DA, Horwitz AF (1996) Cell migration: a physically integrated molecular process. Cell 84(3):359–369
68. David Logan J, Allman ES, Rhodes JA (2005) Mathematical models in biology. Am Math Mon 112(9):847
69. Pengfei L, Weaver VM, Werb Z (2012) The extracellular matrix: a dynamic niche in cancer progression. J Cell Biol 196(4):395–406

70. Madzvamuse A, George UZ (2013) The moving grid finite element method applied to cell movement and deformation. Finite Elem Anal Des 74:76–92
71. Mahadevan S (1997) Monte carlo simulation. Mechanical engineering-New York and Basel-Marcel Dekker-, pp 123–146
72. Massalha S, Weihs D (2016) Metastatic breast cancer cells adhere strongly on varying stiffness substrates, initially without adjusting their morphology. Biomech Model Mechanobiol 16(3):961–970
73. McDougall SR, Anderson ARA, Chaplain MAJ (2006) Mathematical modelling of dynamic adaptive tumour-induced angiogenesis: clinical implications and therapeutic targeting strategies. J Theor Biol 241(3):564–589
74. McDougall SR, Anderson ARA, Chaplain MAJ, Sherratt JA (2002) Mathematical modelling of flow through vascular networks: implications for tumour-induced angiogenesis and chemotherapy strategies. Bull Math Biol 64(4):673–702
75. Ira M, George C, Glenn D (2011) Cancer immunotherapy comes of age. Nature 480(7378):480–489
76. Merkher Y, Weihs D (2017) Proximity of metastatic cells enhances their mechanobiological invasiveness. Ann Biomed Eng 45(6):1399–1406
77. Merks RMH, Koolwijk P (2009) Modeling morphogenesis in silico and in vitro: towards quantitative, predictive, cell-based modeling. Math Model Nat Phenom 4(4):149–171
78. Florian M, Michael B, Petros K (2008) A hybrid model for three-dimensional simulations of sprouting angiogenesis. Biophys J 95(7):3146–3160
79. Mooney CZ (1997) Monte carlo simulation, vol 116. Sage Publications, California
80. Moreira J, Deutsch A (2002) Cellular automaton models of tumor development: a critical review. Adv Complex Syst 5(02n03):247–267
81. Murray JD (2003) Mathematical biology ii: spatial models and biomedical applications, 3rd edn. Springer, Berlin
82. Namazi H, Kulish VV, Wong A (2015) Mathematical modelling and prediction of the effect of chemotherapy on cancer cells. Sci Rep 5:13583
83. Ng MR, Brugge JS (2009) A stiff blow from the stroma: collagen crosslinking drives tumor progression. Cancer cell 16(6):455–457
84. Özdemir BC, Pentcheva-Hoang T, Carstens JL, Zheng X, Chia-Chin W, Simpson TR, Laklai H, Sugimoto H, Kahlert C, Novitskiy SV et al (2014) Depletion of carcinoma-associated fibroblasts and fibrosis induces immunosuppression and accelerates pancreas cancer with reduced survival. Cancer cell 25(6):719–734
85. Paszek MJ, Zahir N, Johnson KR, Lakins JN, Rozenberg GI, Gefen A, Reinhart-King CA, Margulies SS, Dembo M, Boettiger D et al (2005) Tensional homeostasis and the malignant phenotype. Cancer cell 8(3):241–254
86. Powathil G, Kohandel M, Sivaloganathan S, Oza A, Milosevic M (2007) Mathematical modeling of brain tumors: effects of radiotherapy and chemotherapy. Phys Med Biol 52(11):3291
87. Throm Quinlan AM, Sierad LN, Capulli AK, Firstenberg LE, Billiar KL (2011) Combining dynamic stretch and tunable stiffness to probe cell mechanobiology in vitro. PloS one 6(8):e23272
88. Radszuweit M, Block M, Hengstler JG, Schöll E, Drasdo D (2009) Comparing the growth kinetics of cell populations in two and three dimensions. Phys Rev E 79(5):051907
89. Ramis-Conde I, Chaplain MAJ, Anderson ARA (2008) Mathematical modelling of cancer cell invasion of tissue. Math Comput Model 47(5):533–545
90. Reinhardt CA (1994) Alternatives to animal testing: new ways in the biomedical sciences, trends and progress. https://www.cabdirect.org/cabdirect/abstract/19952217321
91. Reinhart-King CA, Dembo M, Hammer DA (2008a) Cell-cell mechanical communication through compliant substrates. Biophys J 95(12):6044–6051
92. Reinhart-King CA, Dembo M, Hammer DA (2008b) Cell-cell mechanical communication through compliant substrates. Biophys J 95(12):6044–6051
93. Rejniak KA (2007) An immersed boundary framework for modelling the growth of individual cells: an application to the early tumour development. J Theor Biol 247(1):186–204

94. Rejniak KA, Anderson ARA (2011) Hybrid models of tumor growth. Wiley Interdiscip Rev Syst Biol Med 3(1):115–125
95. Rejniak KA, Dillon RH (2007) A single cell-based model of the ductal tumour microarchitecture. Comput Math Methods Med 8(1):51–69
96. Lisanne R, Sonja B, Roeland M (2016) Modelling the growth of blood vessels in health and disease. ERCIM News 104:36–37
97. Rhim AD, Oberstein PE, Thomas DH, Mirek ET, Palermo CF, Sastra SA, Dekleva EN, Saunders T, Becerra CP, Tattersall IW et al (2014) Stromal elements act to restrain, rather than support, pancreatic ductal adenocarcinoma. Cancer cell 25(6):735–747
98. Rothman DH, Zaleski S (2004) Lattice-gas cellular automata: simple models of complex hydrodynamics, vol 5. Cambridge University Press, Cambridge
99. Marc D Ryser, Svetlana V Komarova (2015) Mathematical modeling of cancer metastases. Comput Bioeng 211–230
100. Hélène S, Emmanuel D (2012) Within tumors, interactions between t cells and tumor cells are impeded by the extracellular matrix. OncoImmunology 1(6):992–994
101. Neil S (2012) Modelling: computing cancer. Nature 491(7425):S62–S63
102. Gernot S, Michael M-H (2005) Multicellular tumor spheroid in an off-lattice voronoi-delaunay cell model. Phys Rev E 71(5):051910
103. Shirinifard A, Gens JS, Zaitlen BL, Popławski NJ, Swat M, Glazier JA (2009) 3D multi-cell simulation of tumor growth and angiogenesis. PloS one 4(10):e7190
104. Siemann DW (2002) Vascular targeting agents. Horizons in cancer therapeutics: from bench to bedside cancer 3:4–15
105. Simmons A, Burrage PM, Nicolau DV, Lakhani SR, Burrage K (2017) Environmental factors in breast cancer invasion: a mathematical modelling review. Pathology 49(2):172–180
106. Katrin S, Alessandra M (2006) Modeling anticancer drug-dna interactions via mixed qm/mm molecular dynamics simulations. Org Biomol Chem 4(13):2507–2517
107. Stephanou A, McDougall SR, Anderson ARA, Chaplain MAJ (2005) Mathematical modelling of flow in 2d and 3d vascular networks: applications to anti-angiogenic and chemotherapeutic drug strategies. Math Comput Model 41(10):1137–1156
108. András S, Merks Roeland MH (2013) Cellular potts modeling of tumor growth, tumor invasion, and tumor evolution. Front Oncol 3:87
109. Tanaka G, Yoshito Hirata S, Goldenberg L, Bruchovsky N, Aihara K (2010) Mathematical modelling of prostate cancer growth and its application to hormone therapy. Philos Trans R Soc Lond A Math Phys Eng Sci 368(1930):5029–5044
110. Thompson DW et al. (1942) On growth and form. On growth and form
111. Tlsty TD, Coussens LM (2006) Tumor stroma and regulation of cancer development. Annu Rev Pathol Mech Dis 1:119–150
112. Turjanski AG, Gerhard Hummer J, Gutkind S (2009) How mitogen-activated protein kinases recognize and phosphorylate their targets: a qm/mm study. J Am Chem Soc 131(17):6141–6148
113. Turner S, Sherratt JA (2002) Intercellular adhesion and cancer invasion: a discrete simulation using the extended potts model. J Theor Biol 216(1):85–100
114. Jozef VD, Paul P, Jean-Pierre L, Ghislain O (1992) Structural and functional identification of two human, tumor-derived monocyte chemotactic proteins (mcp-2 and mcp-3) belonging to the chemokine family. J Exp Med 176(1):59–65
115. Paul Van Liedekerke MM, Palm NJ, Drasdo D (2015) Simulating tissue mechanics with agent-based models: concepts, perspectives and some novel results. Comput Part Mech 2(4):401–444
116. van Oers RFM, Rens EG, LaValley DJ, Reinhart-King CA, Merks RMH (2014) Mechanical cell-matrix feedback explains pairwise and collective endothelial cell behavior in vitro. PLoS Comput Biol 10(8):e1003774
117. Vermolen FJ (2015) Particle methods to solve modelling problems in wound healing and tumor growth. Comput Part Mech 2(4):381–399
118. Vermolen FJ, Gefen A (2012a) A semi-stochastic cell-based formalism to model the dynamics of migration of cells in colonies. Biomech Model Mechanobiol 11(1):183–195

119. Vermolen FJ, Gefen A (2012b) A semi-stochastic cell-based formalism to model the dynamics of migration of cells in colonies. Biomech Model Mechanobiol 11(1–2):183–195
120. Vermolen FJ, Gefen A (2013a) A phenomenological model for chemico-mechanically induced cell shape changes during migration and cell-cell contacts. Biomech Model Mechanobiol 12(2):301–323
121. Vermolen FJ, Gefen A (2013b) A semi-stochastic cell-based model for in vitro infected 'wound'healing through motility reduction: a simulation study. J Theor Biol 318:68–80
122. Vermolen FJ, Mul MM, Gefen A (2014) Semi-stochastic cell-level computational modeling of the immune system response to bacterial infections and the effects of antibiotics. Biomech Model Mechanobiol 13(4):713–734
123. Vermolen FJ, Van der Meijden RP, Van Es M, Gefen A, Weihs D (2015) Towards a mathematical formalism for semi-stochastic cell-level computational modeling of tumor initiation. Ann Biomed Eng 43(7):1680–1694
124. Vincent TL, Gatenby RA (2008) An evolutionary model for initiation, promotion, and progression in carcinogenesis. Int J Oncol 32(4):729–737
125. Wang James HC, Jeen-Shang L (2007) Cell traction force and measurement methods. Biomech Model Mechanobiol 6(6):361–371
126. Wang Z, Zhang L, Sagotsky J, Deisboeck TS (2007) Simulating non-small cell lung cancer with a multiscale agent-based model. Theor Biol Med Model 4(1):50
127. Ward JP, King JR (1997) Mathematical modelling of avascular-tumour growth. Theor Math Med Biol J IMA and Med Model 14(1):39–69
128. Weens W (2012) Mathematical modeling of liver tumor. PhD thesis, Université Pierre et Marie Curie-Paris VI
129. Winer JP, Chopra A, Kresh JY, Janmey PA (2011) Mechanobiology of cell–cell and cell–matrix interactions. Chapter 2
130. Wolf K, Wu YI, Liu Y, Geiger J, Tam E, Christopher Overall M, Stack S, Friedl P (2007) Multi-step pericellular proteolysis controls the transition from individual to collective cancer cell invasion. Nat Cell Biol 9(8):893
131. Yang L, Witten TM, Pidaparti RM (2013) A biomechanical model of wound contraction and scar formation. J Theor Biol 332:228–248

Estimation of 6 Degrees-of-Freedom Accelerations from Head Impact Telemetry System Outputs for Computational Modeling

Logan E. Miller, Jillian E. Urban and Joel D. Stitzel

Abstract To understand the biomechanical basis of head impacts, finite element (FE) modeling is used to estimate the response throughout the brain in various impact conditions. FE simulation of head motion requires a complete description of kinematics, such as six degrees of freedom (6DOF) linear and rotational acceleration curves defining the boundary conditions. These are not available from many common head impact sensors such as the Head Impact Telemetry (HIT) System. At the same time, there are hundreds of thousands of impacts, likely millions of impacts, collected by HITS which represent an underutilized resource for computational modeling. The goal of this study was to develop an algorithm to determine 6DOF acceleration curves based on the corresponding HITS output data for use in FE modeling. The transformation algorithm was developed from a dataset of 14,767 head impacts collected with the HIT System and the corresponding 6DOF information provided by a published algorithm for this study. The impacts were sorted into impact regions and classified by the polarity of peak accelerations, and characteristic curves for each polarity combination were calculated. The algorithm was validated against 50 random impacts by comparing predicted and true acceleration curves using an objective curve comparison metric, CORA, to quantify error. These results demonstrate the algorithm accurately estimates 6DOF motion characteristics from 5DOF inputs sufficient for the purpose of performing basic biomechanical analyses of the impacts through FE modeling.

1 Introduction

There are approximately 5 million athletes playing organized football in the United States; 2,000 professional players, 100,000 college players, 1.3 million high school players, and 3.5 million youth players [1–3]. Sport-related traumatic brain injury (TBI) is an important public health concern due to the number of people affected

L. E. Miller (✉) · J. E. Urban · J. D. Stitzel
Wake Forest School of Medicine, Winston-Salem, NC, USA
e-mail: logmille@wakehealth.edu

© Springer Nature Switzerland AG 2019
J. M. R. S. Tavares and P. R. Fernandes (eds.), *New Developments on Computational Methods and Imaging in Biomechanics and Biomedical Engineering*, Lecture Notes in Computational Vision and Biomechanics 33, https://doi.org/10.1007/978-3-030-23073-9_8

and unknown and potentially serious resulting conditions. Although football has a high rate of concussion, exposure to repetitive subconcussive head impacts, which occur as part of normal participation in the sport, and associated changes in the brain related to neurodegenerative diseases is of increasing concern [4–8]. The advent of head-sensing technology has allowed researchers to collect real-world on-field head impact data by instrumenting athletes during typical play over the course of entire seasons [9–13]. In this context, head impact exposure in football has been extensively studied; however, the biomechanical basis of subconcussive head impacts is still not well-understood. To better understand the effects of repetitive subconcussive impacts, biomechanical factors of head impact, such as impact location and direction, as well as brain parenchymal deformations can be studied in detail using finite element (FE) models.

Various FE-based studies have been conducted to quantify the strain response of the brain under conditions representative of typical football impacts. In 2014, Ji et al. used the Dartmouth Head Injury Model (DHIM) and the Simulated Injury Monitor (SIMon) to investigate brain-strain-related responses in a range of loading conditions representative football impacts experienced at the youth, high school, and collegiate levels [14]. Brain deformation was measured using deformation metrics proposed to have a correlation to brain injury, such as maximum principal strain (MPS) and von Mises stress [15, 16]. This study also investigated the relative contributions of linear and angular acceleration to the strain response and found that isolated linear acceleration generates negligible strain. A similar study used the DHIM, SIMon, and Wayne State University Brain Injury Model (WSUBIM) models to study regional brain response in the cerebrum, cerebellum, brainstem, and whole brain [17]. Smith et al. (2015) used the UCDBTM to evaluate strain response for indirect, direct, and combined loading scenarios [18]. Darling et al. (2016) used the head model from the Global Human Body Models Consortium (GHBMC) full body model to evaluate the strain response to two typical loading conditions experienced in football—frontal impact and crown impact [19].

Combining the large existing data set of real-world head impacts collected over the years with validated FE brain models presents a valuable opportunity to advance our knowledge of the brain's response to head impact. This is not a straightforward task, when considering the data head-sensing devices provide and the data necessary for FE modeling. FE simulation of head motion requires a full description of the kinematics of the skull. One way to achieve this is through the provision of six degrees of freedom (6DOF) linear and rotational acceleration curves to define the boundary conditions, which is not available from most common head impact sensors. For example, outputs from the Head Impact Telemetry (HIT) System are limited to peak XYZ linear acceleration values, peak XY rotational acceleration values, a 40 ms linear resultant time trace, and an azimuth and elevation angle for each impact [20]. Beckwith et al. published an algorithm, which can be used to estimate 6DOF information from the original accelerations collected by the single axis accelerometers used in the HITS. However, the raw acceleration data for each impact is not provided to researchers using the system under any circumstance and the 6DOF algorithm is only used to generate 6DOF data which is provided to researchers with a collaborative contract or grant with Simbex (Lebanon, NH). This means that a very large percentage of the HITS data collected in the research environment cannot be used for modeling, including for many studies where there is imaging data also available [21–26].

Therefore, the objective of this study was to develop a transformation algorithm to determine 6DOF acceleration curves based on the corresponding HITS output data for use in future finite element studies.

2 Methods

The current study develops a transformation algorithm to estimate complete kinematic descriptions for head impacts from 5DOF HITS output data. The HITS output data include only the following kinematics: resultant linear acceleration time history, peak XYZ linear acceleration values, and peak XY rotational acceleration values. Figure 1 shows complete kinematics for an example head impact; the HITS output is shown in black, whereas the missing information is shown in gray. Thus, the kinematic information displayed in gray is the information that will be produced by the transformation algorithm.

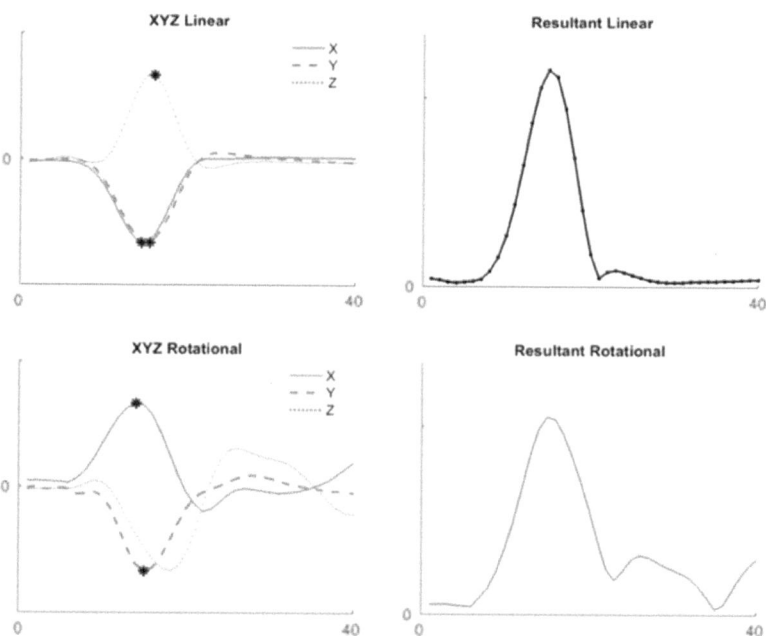

Fig. 1 Head kinematics for an example impact showing information included in the HITS outputs (displayed in black) and the missing information to complete the 6DOF kinematics (displayed in gray)

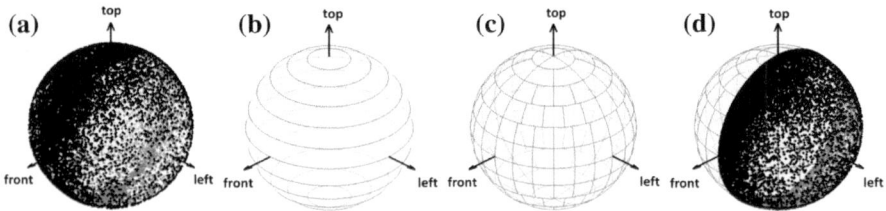

Fig. 2 All impacts plotted on a sphere representing the head (**a**), impact levels (**b**), impact regions (**c**), and all impacts reflected to the left side (**d**)

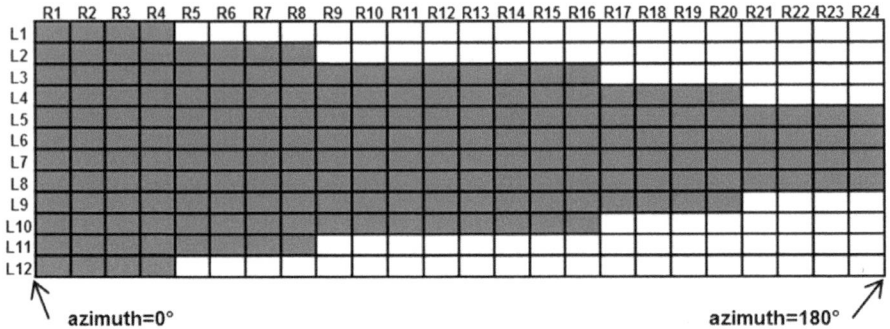

Fig. 3 Numbering scheme for impact regions

The transformation algorithm was developed from an existing dataset of head impacts collected with the HIT System. First, the impacts were sorted into impact regions defined by approximately equal divisions of azimuth and elevation. Twelve impact levels representing 15° increments in elevation angles were defined (Fig. 2b), numbered from the top of the head (1) to the bottom (12). Starting at the impact level with the largest surface area (level 6), regions were divided into 15° azimuth regions. Then, for impact levels 1 through 5, regions were defined using azimuth angles that resulted in the surface area of impact regions closest to the surface area for impact regions from impact level 6. This resulted in between 4 and 24 regions defined per level to define ensure similar surface area impact regions in a spherical coordinate system. This process resulted in a total of 192 impact regions (Fig. 2c).

Next, assuming symmetry, impacts that occurred on the right side of the head were reflected to the left side so all impacts occurred on the left hemisphere (Fig. 2d). To reflect an impact, the azimuth angle was inverted, as well as the following acceleration components: linear Y, rotational X, and rotational Z. A two-dimensional representation of the numbering scheme for impact regions is shown in Fig. 3, which may be considered analogous to a map cartographic projection.

Impacts were then classified by the polarity of peak accelerations with a 1×6 vector of positive or negative ones corresponding to the polarity of XYZ linear and rotational acceleration. For example, the polarity characterization corresponding to the impact is shown in Fig. 4 is $[-1 \; -1 \; -1 \; +1 \; -1 \; -1]$. Impacts corresponding to

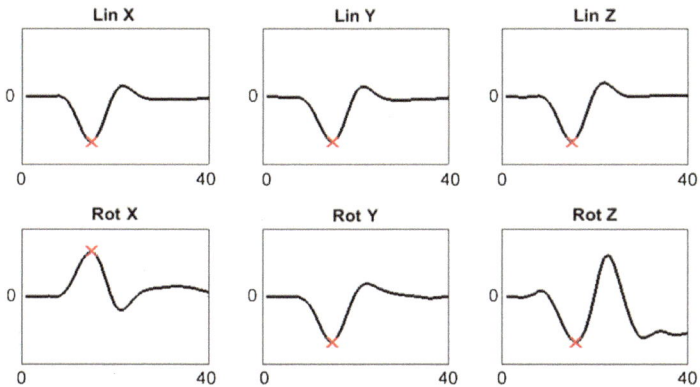

Fig. 4 Example impact demonstrating polarity characterization

Table 1 Unique polarity combinations for level 2, region 2

Lin X	Lin Y	Lin Z	Rot X	Rot Y	Rot Z	# Impacts
−1	−1	−1	+1	−1	−1	123
−1	−1	−1	+1	+1	−1	87
−1	−1	−1	+1	−1	+1	70
+1	−1	−1	+1	+1	−1	9
−1	−1	−1	+1	+1	+1	7
+1	−1	−1	+1	+1	+1	4
+1	+1	+1	−1	+1	+1	4
+1	−1	−1	+1	−1	−1	3
+1	−1	−1	+1	−1	+1	3
−1	−1	−1	−1	−1	−1	1
−1	+1	−1	−1	−1	+1	1
−1	+1	−1	+1	+1	+1	1
−1	+1	+1	−1	+1	−1	1
+1	−1	−1	−1	+1	−1	1
+1	+1	−1	−1	+1	+1	1
+1	+1	+1	−1	−1	+1	1
+1	+1	+1	−1	+1	−1	1

each region were then grouped by unique polarity combinations. For example, there were 17 unique polarity combinations corresponding to the 318 impacts in impact Level 2, Region 2 (Table 1).

To use the algorithm to estimate 6DOF curves for a given HITS impact, characteristic curves corresponding to the appropriate impact region and polarity are determined and then scaled to the peak values output by the HIT System.

To validate the algorithm, 50 random impacts from the dataset were selected and the true and predicted acceleration curves were compared. CORA, an objective comparison metric, was used to quantify error between the true and predicted curves [27]. CORA is an objective rating method combining two independent submethods, a corridor rating, and a cross-correlation rating. These two ratings range from 0 to 1 and are averaged to determine the CORA rating (1 indicates a perfect match). The corridor method computes a rating based on where the simulation curve falls in relation to corridors around the experimental curve. The cross-correlation method is based on ratings for the phase shift, size, and shape of time-shifted curves. In addition to incorporating both point-by-point and peak value comparisons for assessing model performance, CORA is also able to evaluate the cross-correlation of two curves. CORA scores were calculated for all six acceleration curves and averaged to get a single rating for each tested impact.

3   Results

The dataset consisted of 14,767 impacts which were sorted into 192 impact regions. The number of impacts associated with an individual impact region ranged from 9 to 710. A total of 8,060 (54.6%) impacts were reflected from the right hemisphere to the left. The number of unique polarities per impact region ranged from 4 to 44 combinations.

Characteristic curves for each unique polarity combination were calculated by averaging aligned normalized acceleration curves. The characteristic curves for the impact Level 4, Region 2 (Fig. 5) are shown in Fig. 6. 6DOF curves were generated for each impact by scaling the characteristic curves to the peak values output by the HIT System given the impact region and polarity.

To validate the characteristic curves produced by the 5DOF to 6DOF algorithm, 50 random impacts were selected and the curves predicted by the algorithm were compared to the true acceleration curves for that impact. CORA scores were calculated for all six acceleration curves and averaged to compute a single rating for each tested impact. The mean, minimum, and maximum CORA scores of the 50

Fig. 5 Impacts associated
with impact level 4, region 2

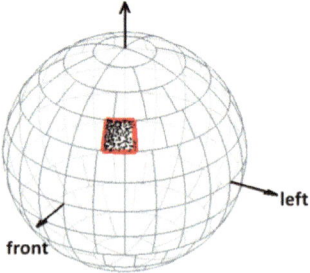

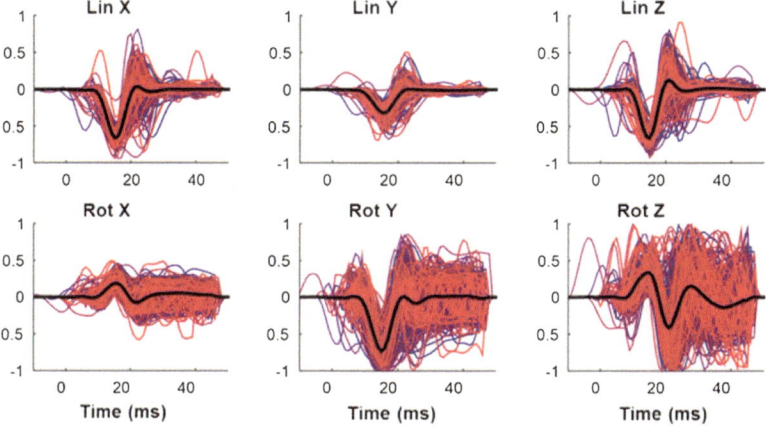

Fig. 6 Characteristic curves for impact region highlighted in Fig. 5

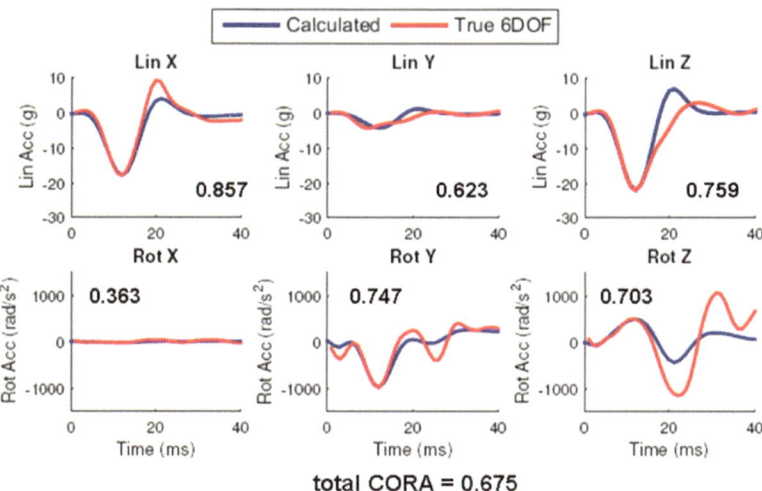

Fig. 7 Validation of an example sampled impact

validation impacts were 0.497, 0.267, and 0.733, respectively. Comparison of true and predicted curves for an example sampled impact is shown in Fig. 7, which had an average CORA score of 0.675.

4 Discussion

The algorithm presented in this study utilizes calculated characteristic curves associated with specific polarities at each impact region to compute 6DOF data from 5DOF

HITS data. In this approach, the impact region and polarity combination is calculated to determine the associated characteristic curves. The curves are then scaled to the peak values of the acceleration components determined from the HITS output. This algorithm was validated against 50 random impacts and resulted in mean, minimum, and maximum CORA scores of the 50 validation impacts were 0.497, 0.267, and 0.733, respectively. These results demonstrate the algorithm accurately estimates 6DOF motion characteristics from 5DOF inputs sufficient for the purpose of performing basic biomechanical analyses of the impacts through FE modeling. This algorithm allows the leveraging of hundreds of thousands of head impacts collected using the HITS system, many with accompanying medical imaging and neurocognitive data, over years of research to be further studied using FE brain models. This ability will contribute to the goal of quantifying subconcussive impact exposure, and perhaps elucidating concussion mechanisms, and identifying finite element based metrics discriminating concussion injury thresholds.

A limitation of this approach is that the use of estimated inputs to the algorithm (FE boundary conditions) to predict strain and other metrics correlated with an injury. While the algorithm provides promising results and demonstrates the ability to closely predict acceleration curves, small differences may still result in differences in the FE output.

5 Conclusion

The goal of this study was to develop a transformation algorithm to determine 6DOF acceleration curves based on the corresponding HITS output data for use in future FE studies. An algorithm consisting of a set of characteristic curves was calculated which can be used with HITS output values to produce estimated 6DOF acceleration curves. These results demonstrate the algorithm accurately estimates 6DOF motion characteristics from 5DOF inputs to the degree necessary for using in FE simulation.

References

1. Guskiewicz KM, Weaver NL, Padua DA, Garrett WE (2000) Epidemiology of concussion in collegiate and high school football players. Am J Sport Medicine 28:643–650
2. Powell JW, Barber-Foss KD (1999) Traumatic brain injury in high school athletes. JAMA 282:958–963
3. Daniel RW, Rowson S, Duma SM (2012) Head impact exposure in youth football. Ann Biomed Eng 40:976–981
4. Stamm JM, Koerte IK, Muehlmann M, Pasternak O, Bourlas AP, Baugh CM et al (2015) Age at first exposure to football is associated with altered corpus callosum white matter microstructure in former professional football players. J Neurotrauma 32:1768–1776
5. Stamm JM, Bourlas AP, Baugh CM, Fritts NG, Daneshvar DH, Martin BM et al (2015) Age of first exposure to football and later-life cognitive impairment in former NFL players. Neurology 84:1114–1120

6. Stern RA, Riley DO, Daneshvar DH, Nowinski CJ, Cantu RC, McKee AC (2011) Long-term consequences of repetitive brain trauma: chronic traumatic encephalopathy. Pm&r 3:S460–S467

7. Montenigro PH, Alosco ML, Martin B, Daneshvar DH, Mez J, Chaisson C, et al (2016) Cumulative head impact exposure predicts later-life depression, apathy, executive dysfunction, and cognitive impairment in former high school and college football players. J Neurotrauma

8. Alosco ML, Tripodis Y, Jarnagin J, Baugh CM, Martin B, Chaisson CE et al (2017) Repetitive head impact exposure and later-life plasma total tau in former national football league players. Alzheimer's Dement: Diagn Assess Dis Monitoring 7:33–40

9. Broglio SP, Sosnoff JJ, Shin S, He X, Alcaraz C, Zimmerman J (2009) Head impacts during high school football: a biomechanical assessment. J Athl Train 44:342

10. Duma SM, Manoogian SJ, Bussone WR, Brolinson PG, Goforth MW, Donnenwerth JJ et al (2005) Analysis of real-time head accelerations in collegiate football players. Clin J Sport Medicine 15:3–8

11. Urban JE, Davenport EM, Golman AJ, Maldjian JA, Whitlow CT, Powers AK et al (2013) Head impact exposure in youth football: high school ages 14 to 18 years and cumulative impact analysis. Ann Biomed Eng 41:2474–2487

12. Kelley ME, Urban JE, Miller LE, Jones DA, Espeland MA, Davenport EM et al (2017) Head impact exposure in youth football: comparing age and weight based levels of play. J Neurotrauma

13. Cobb BR, Urban JE, Davenport EM, Rowson S, Duma SM, Maldjian JA et al (2013) Head impact exposure in youth football: elementary school ages 9–12 years and the effect of practice structure. Ann Biomed Eng 41:2463–2473

14. Ji S, Zhao W, Li Z, McAllister TW (2014) Head impact accelerations for brain strain-related responses in contact sports: a model-based investigation. Biomech Model Mechanobiol 13:1121–1136

15. Zhang L, Yang KH, King AI (2004) A proposed injury threshold for mild traumatic brain injury. J Biomech Eng 126:226–236

16. Kleiven S (2007) Predictors for traumatic brain injuries evaluated through accident reconstructions. Stapp Car Crash J 51:81–114

17. Ji S, Ghadyani H, Bolander RP, Beckwith JG, Ford JC, McAllister TW et al (2014) Parametric comparisons of intracranial mechanical responses from three validated finite element models of the human head. Ann Biomed Eng 42:11–24

18. Smith TA, Halstead PD, McCalley E, Kebschull SA, Halstead S, Killeffer J (2015) Angular head motion with and without head contact: implications for brain injury. Sport Eng 18:165–175

19. Darling T, Muthuswamy J, Rajan S (2016) Finite element modeling of human brain response to football helmet impacts. Comput Methods Biomech Biomed Eng, pp 1–11

20. Beckwith JG, Greenwald RM, Chu JJ (2012) Measuring head kinematics in football: correlation between the head impact telemetry system and Hybrid III headform. Ann Biomed Eng 40:237–248

21. Davenport EM, Whitlow CT, Urban JE, Espeland MA, Jung Y, Rosenbaum DA et al (2014) Abnormal white matter integrity related to head impact exposure in a season of high school varsity football. J Neurotrauma 31:1617–1624

22. Bahrami N, Sharma D, Rosenthal S, Davenport EM, Urban JE, Wagner B et al (2016) Subconcussive head impact exposure and white matter tract changes over a single season of youth football. Radiology 281:919–926

23. Davenport EM, Apkarian K, Whitlow CT, Urban JE, Jensen JH, Szuch E et al (2016) Abnormalities in diffusional kurtosis metrics related to head impact exposure in a season of high school varsity football. J Neurotrauma 33:2133–2146

24. Nencka AS, Meier TB, Wang Y, Muftuler LT, Wu Y-C, Saykin AJ, et al (2017) Stability of MRI metrics in the advanced research core of the NCAA-DoD concussion assessment, research and education (CARE) consortium. Brain Imaging Behav [Internet]. [cited 2018 Jul 15]; Available from: http://link.springer.com/10.1007/s11682-017-9775-y

25. Mustafi SM, Harezlak J, Koch KM, Nencka AS, Meier T, West JD, et al (2017) Acute white-matter abnormalities in sports-related concussion: a diffusion tensor imaging study from the NCAA-DoD CARE consortium. J Neurotrauma [Internet]. [cited 2018 Jul 15]; Available from: http://www.liebertpub.com/doi/10.1089/neu.2017.5158

26. Rowson S, Duma SM, Stemper BD, Shah A, Mihalik JP, Harezlak J et al (2018) Correlation of concussion symptom profile with head impact biomechanics: a case for individual-specific injury tolerance. J Neurotrauma 35:681–690

27. Gehre C, Gades H, Wernicke P (2009) Objective rating of signals using test and simulation responses. Paper presented at: 21st ESV conference

Physiological Cybernetics: Methods and Applications

Daniela Iacoviello

Abstract In this paper, it is discussed how physiological systems can be regulated by using the control theory as well as methodologies of system analysis, modeling, and identification. In physiology, the natural tendency to homeostasis, despite changes in the environments, implies a feedback-control scheme. The study of the natural regulation in physiological systems could help in its replacing when pathological situations are present. The basic concepts of homeostasis, modeling and control are here recalled, and some case studies are described.

1 Introduction: Physiological Cybernetics

The term "physiological cybernetics", introduced by Wiener in 1948, refers to the possibility of determining models able to suitably describe physiological systems in the framework of systems and control theory.

The study of physiological phenomena in the control context implies many difficulties [32]; first, the definition of the state variables, the inputs and the outputs are not always evident. This aspect is strictly related with the modeling of the phenomena under exam and with the fact that the system to be controlled is a controller itself; physiological systems are usually unknown and different systems interact between each other, thus making difficult and complex the description of the phenomena. They are systems intrinsically time variant and highly nonlinear, and linearization does not provide, usually, realistic results. When a model is defined according to the physical relations, its parameters must be determined generally considering measures taken in the correspondent real conditions, with in vivo experiments rather difficult to be reproduced and sometimes expensive. The control designed for a physiological system requires versatility and the capability to face different functions; this con-

D. Iacoviello (✉)
Department of Computer, Control and Management Engineering Antonio Ruberti, Sapienza University of Rome, Via Ariosto 25, 00185 Rome, Italy
e-mail: daniela.iacoviello@uniroma1.it

© Springer Nature Switzerland AG 2019
J. M. R. S. Tavares and P. R. Fernandes (eds.), *New Developments on Computational Methods and Imaging in Biomechanics and Biomedical Engineering*, Lecture Notes in Computational Vision and Biomechanics 33, https://doi.org/10.1007/978-3-030-23073-9_9

trol system should act not only with a feedback action but also assuming that the controller characteristics could change in an adaptive way.

The specificity of dealing with physiological control systems requires researchers able to interact with experts of the different disciplines involved, physiology, modeling, identification, measurements, data analysis and, of course, automatic control. The latter may require techniques in the framework of nonlinear and/or optimal and/or adaptive control to deal with the complexity of this kind of phenomena. The importance of communication between researchers with different skills was stated by Wiener [46], who, referring specifically to mathematicians and physiologists, emphasized the importance of the researchers involved in interdisciplinary studies to share different expertise and knowledge. This difficulty is enhanced also in [12], where it is noted that generally control theory has been applied in electrical, mechanical and aerospace engineering field, but rarely in bioengineering, despite the fact, as will be shown in this paper, that the basic elements of automatic control, such as the feedback, are intrinsic in physiological systems.

The availability of a large amount of physiological and environmental data and of new techniques for combining multiple data sources have further increased the interest to physiological cybernetics, thus allowing the effective action and improvement of the research activities. This is also due to the undeniable advantage of the simulation aspect to study the behaviors of the variables and the effects of possible actions, *before* acting on a real system or, even better, *along* with experiments in vivo. In fact, the interaction of the modeling, simulation and control with experiments improves the knowledge of the phenomena, suggesting possible control strategies.

The applications of automatic control actions to physiological problems includes the following studies: diabetes [11, 22], pharmacokinetics/pharmacodynamics [3, 33] and thyroid control [41, 45] allowing the drug dosage determination, epidemic analysis and control [4, 35, 29], respiratory and cardiac modeling and control [9, 31], bone remodeling [1, 5, 43], gene network regulation [14], rehabilitation devices [37, 38], just to mention few examples. All these topics may be appreciated also for their interdisciplinary characteristics, including in the term "physiological systems" also compartmental models describing interactions between individuals. The fields of applications are in a wide range, in different contexts and from various points of view.

This paper is organized as follows; in Sect. 2 the concept of homeostasis is discussed, and some examples are recalled. In Sect. 3, the modeling and control peculiarities in physiological cybernetics are described pointing out, whenever necessary, the differences with respect to other application fields. In Sect. 4 some case studies of control applied to physiological systems are discussed; they refer to glucose regulation, epidemic modeling control, bone remodeling, dosage determination in a chemotherapy and pupil light reflex.

2 Homeostasis

The concept of homeostasis dates to 1926, when Cannon introduced the word "homeostasis" to describe the stability of various constituents in body fluids and of core temperature [6]; in particular, he wrote: "When a factor is known which can shift a homeostatic state in one direction it is reasonable to look for automatic control of that factor or for a factor or factors having an opposing effect." Physiological functions require stability conditions despite changes in the environments, Fig. 1. Homeostasis is the regulation and maintenance of the internal environment of the body; the body must remain within a narrow range of conditions, like, for example, the body temperature.

The homeostasis, or equilibrium state, is possible by suitable feedback action; it acts at different levels, molecular, cellular, organism, and population [27]. More precisely, at the molecular level, it limits the final quantity of products due to the enzymatic system; at the cellular level, it limits the mitotic process in a cellular population when the cells become too numerous; at the organism level, the various mechanisms work together in different modalities; at population level it regulates the flux of subjects from one class to the other. The homeostatic system is based on the following principal items that, together, represent the feedback response, Fig. 2:

- the stimulus: it is the change that stimulates the receptor to activate the regulation;
- the receptor: it gets the external and internal conditions;
- the control center: it compares the condition identified by the receptor and the optimal one and acts consequently;
- the actuator: it applies the decisions of the control center.

In the classical control language, the corresponding block diagram scheme is shown in Fig. 3.

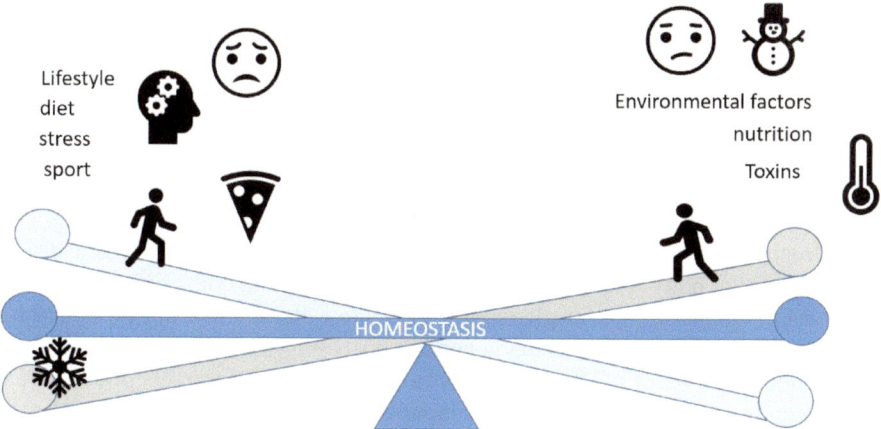

Fig. 1 Non-homeostatic situation as a condition for disease

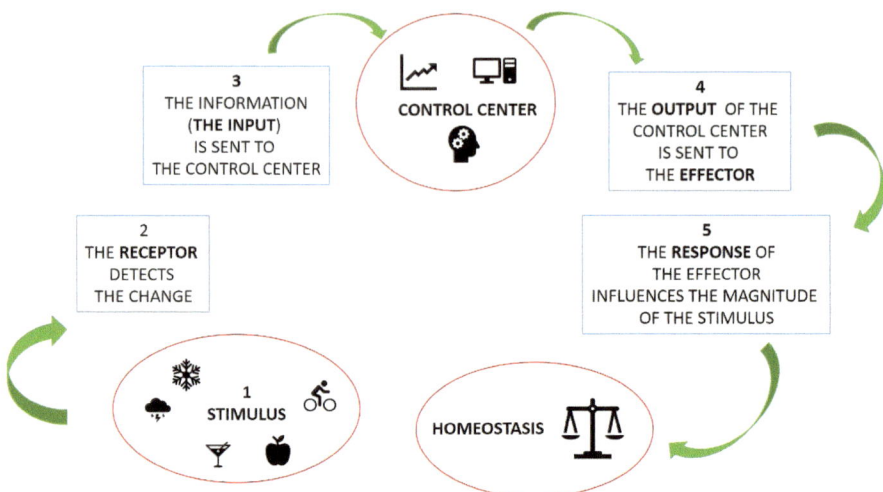

Fig. 2 The feedback response as the homeostatic scheme

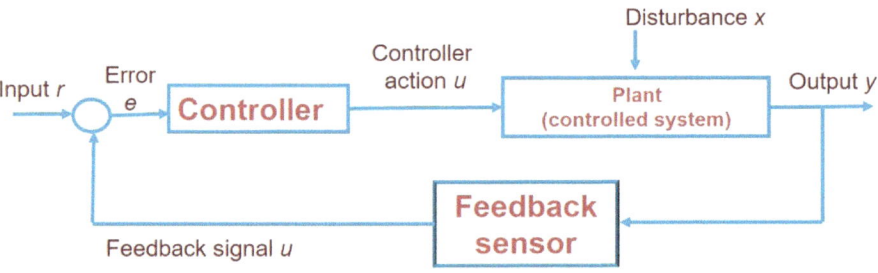

Fig. 3 A classical feedback scheme

As said, the homeostasis guarantees the maintenance of almost constant physiological parameters despite changes in the external and internal conditions. It is so important that a redundancy in the control mechanism is usually present: sometimes for a single "variable" more than one control system is present, and more than one effector can act; redundancy guarantees stability of the variable despite different perturbations. Examples of redundancy are in blood pressure and in the temperature control mechanisms.

The regulation of the body temperature is a typical example of homeostasis, see Fig. 4. When the body temperature falls, all the body acts to conserve the heat: the blood vessels constrict, sweat glands do not secrete fluid, shivering produced by involuntary contraction of muscles generates heats. On the other hand, when the body temperature arises all the body acts to lose heat: blood vessels dilate, sweat glands secrete fluids, and, as the fluids evaporate, heat is lost.

Another classic example of homeostasis regards glucose regulation, Fig. 5. If the level of glucose is high (for example, after a meal), the beta cells of pancreas release

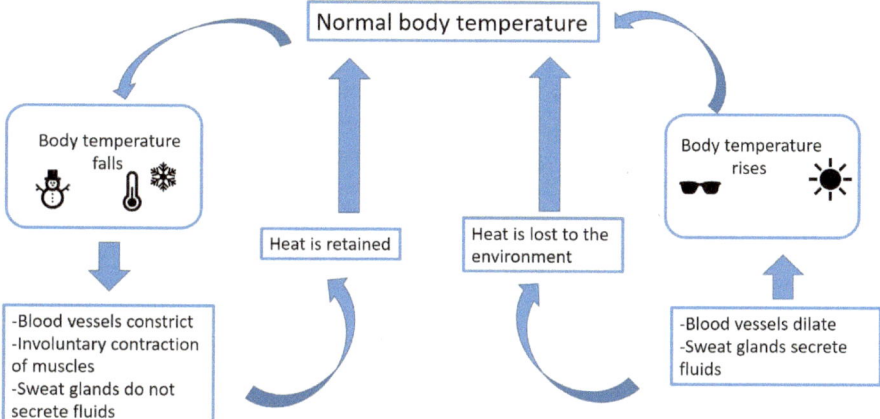

Fig. 4 Body temperature homeostasis cycle

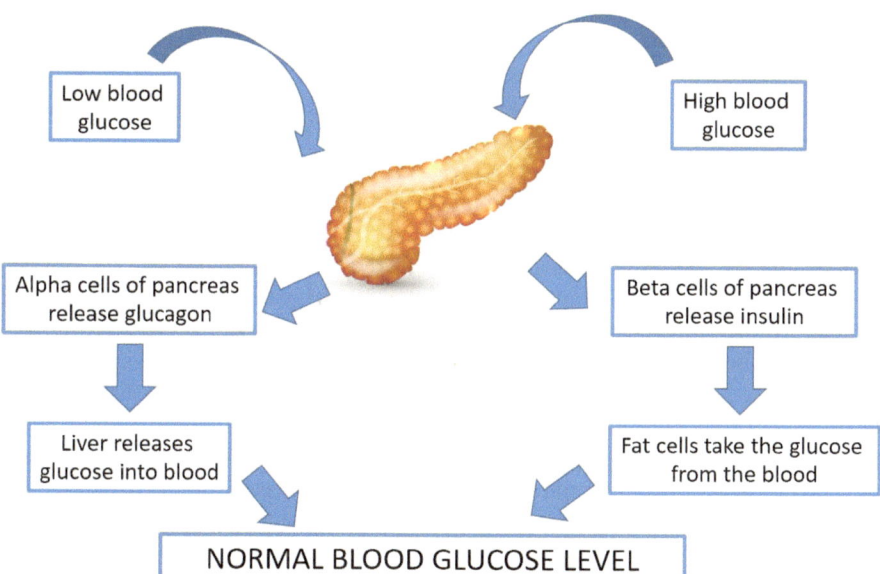

Fig. 5 Glucose homeostasis cycle

insulin, that acts as a "key" allowing the fat cells to take glucose from the blood: the normal blood glucose level is thus achieved. On the contrary, if the blood glucose level is low, the alpha cells of pancreas release glucagon and the liver releases glucose into the blood, thus allowing the blood to achieve the normal glucose level.

3 Modeling and Control

In this section, the modeling and control peculiarities in physiological cybernetics
are discussed pointing out, whenever necessary, the differences with respect to other
application fields. The complete replacement of natural control action by an external automatic method is not an easy task; it requires the knowledge (not completely
available) of the phenomenon, biomedical instruments and devices, data analysis and
processing, modeling and control. The possibility of a mathematical representation
could allow a better understanding of the relations among the variables of interest; moreover, control strategies could be implemented to mimic the natural control
mechanisms, thus allowing the determination of possible action (for example the
medication) whenever required.

In the following, there will be discussed the noteworthy aspects of modeling and
control from the physiological cybernetics point of view.

Modeling Aspects
To face a physiological control problem, the knowledge of the underlying medical
and, if necessary, also social aspects is needed to guide the modeling and control
phases, [8]. In Sect. 4, some case studies are discussed enhancing the rationale under
the choices regarding the description and the control actions adopted. As far as the
modeling point of view, two main approaches may be cited: the data-driven models
and the system modeling; in the first case, the models are based on experimental
data, looking for quantitative description of the phenomena. This kind of models
are particularly appropriate when there is a lack of knowledge of the physiological
mechanisms involved: sometimes they are referred to as "black box models". When
it is possible to get a priori knowledge of the phenomena, some assumptions are made
and, depending on the degree of approximation chosen, a model is proposed. In this
framework, one can deal with linear/nonlinear models, of distributed, or stochastic,
or discrete modeling and so on; all of these represent approximations that must be
considered. Another common aspect in the modeling phase is tentative to use well-
known models to describe physiological phenomena; for example, electrical models
could help in describing the respiratory mechanics, as well as the theory of oscillators,
and in general of periodic systems, are useful to describe the cardiac output and the
periodic breathing [25, 31].

Once a first choice of the model has been proposed, simulations are useful to
examine the behaviors of the variables and therefore to check if the chosen representation is appropriate, comparing the output of the model with the real data and/or
the information available on the physiology. This problem involves also the identification of the models' parameters: data are required, along with the proper choice
of the experiments design. Finally, model validation studies whether the model is
adequate with the purposes; it is not "the last step", since during the overall modeling
procedure one or more phase could be repeated to improve and/or to simplify the
model as well as to update the numerical values of the parameters.

Control Aspects

All the physiological aspects of our body may be described in terms of "maintenance of homeostatic conditions" involving feedback-control scheme, both negative and positive ones, as shown in Fig. 3. Both schemes are common in nature; in the former case, the negative feedback allows the system to act as regulator, thus reducing the error signal. The two cases previously proposed, the regulation of the body temperature and of the glucose, are obviously examples of negative feedback schemes, since the aim of the control action is to restore the physiological conditions, despite the variation due, for example, to the external changes of temperature and to the assumption of food, respectively. Another example of negative feedback is the pupil light reflex; the size of the pupil depends on muscles activated or left to relax: they receive control signals from the brain, depending on the light level. Pupil size depends on the interaction of the parasympathetic and the sympathetic nervous system, controlled by the central nervous one. When the light level is high, there is the activation of constricting muscles which shrink the pupil area thus decreasing the light flux on the retina. On the other hand, when the light level is low, the smooth cells of the radial muscle contract and the result is the dilation of the pupil.

In the positive feedback scheme, the feedback signal is added to the input, rather than being subtracted, thus leading to a *vicious cycle of events* [31]. A typical example of positive feedback regards the oxytocin cycle and the birth of a baby. The baby pushes against the cervix causing its stretching; the latter causes nerve impulses to be sent to the brain that stimulates the pituitary gland to release oxytocin; it increases the uterus contractions and the cycle goes on up to the birth of the baby.

The importance of the study of physiological systems and their control mechanisms is summarized in Fig. 6; more precisely, in Fig. 6a it is shown the feedback scheme of the glucose–insulin interaction. When the action of the beta-cells falls, an external control must be applied, Fig. 6b: it must try to replicate *automatically* the spontaneous insulin production and action. Again, this implies the involvement of many different disciplines, such as modeling, instrumentation, and feedback-control strategies.

As already noted, models of physiological systems are in general highly nonlinear and the control methods to be used must face this difficulty. Some important physiological problems, such as the study of the cardiac output, the fluctuation of ventilation and the circadian model, and their pathologies, may be adequately described by using oscillator modeling along with the phase-plane analysis and the isocline method. The nonlinearity characteristics of the physiological systems may lead to complex patterns that can be efficiently described within the theory of bifurcation [47], as in the logistic equation where fractal structures are present in the time evolution of the solution. Also in physiological control systems linearization is possible and sometimes used, for example with the theory of small signal perturbation; an interesting example of this approximation is used in the model of the Cheyne–Stokes breathing [7, 36].

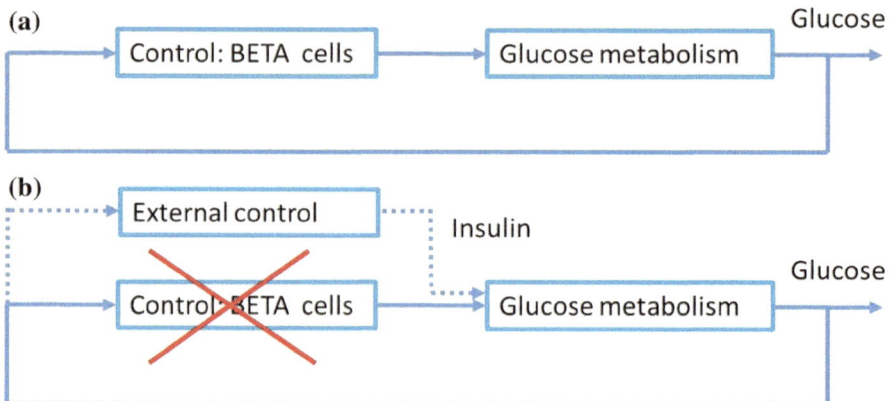

Fig. 6 Glucose-insulin feedback scheme; **a** normal case; **b** in a diabetic patient an external control is required

4 Case Studies

In this Section, there will be discussed some examples of the application of automatic control theory to real problems, such as the regulation of glucose, the dynamic of population, bone remodeling, the chemotherapeutic strategies definition, and the pupil light reflex.

The examples proposed present the dynamics and interaction that could be described at different levels; in all the cases it has been chosen a complete but not too complicated modeling to focus the attention on the homeostatic aspects and modeling motivation, other than the regulation. As far as the control point of view, many results are available in the literature and could substitute the ones discussed herein, provided the compatibility of the variables used: this is an interesting modular characteristic that guarantees versatilities both of the modeling and of the control steps.

The examples are discussed in a "constructive way", trying to involve the reader in the motivations of the choices. After a short medical and physiological introduction useful to understand the problem, the mathematical description is briefly proposed referring to recently published papers of different authors.

The case studies discussed are heterogeneous and may be referred to as "physiological control systems" broadly speaking.

Regulation of Glucose

Diabetes is a chronic disease that occurs either when the pancreas does not produce enough insulin (type 1) or when the body cannot effectively use the produced one (type 2). Insulin is the hormone that regulates blood sugar. The common effect of uncontrolled diabetes is the raised blood sugar (hyperglycemia) that leads to serious damage, like in blood vessels and nerves. The literature on the regulation of glucose is vast, from one of the first model by Stolwijk and Hardy of [40], to the Hovorka one of [26], up to the more recent models in [15, 30]. In this Section, it has been chosen to discuss the approach of [39] in which the simple model of Stolwijk and Hardy is assumed, and an action based on a PID control is introduced.

Denoting by $x(t)$ and $y(t)$, the glucose and the insulin quantities respectively, their dynamics are described by the following nonlinear equations:

$$
\begin{array}{ll}
C_G \frac{dx}{dt} = U(t) + Q_L - \lambda x - \nu xy & x \le \theta \\
C_G \frac{dx}{dt} = U(t) + Q_L - \lambda x - \nu xy - \mu(x - \theta) & x > \theta \\
C_I \frac{dy}{dt} = U_I(t) - \alpha y & x \le \varphi \\
C_I \frac{dy}{dt} = U_I(t) - \alpha y + \beta(x - \varphi) & x > \varphi
\end{array}
$$

where θ and φ are thresholds that, when exceeded, imply the renal loss and the production of the insulin, respectively; $U(t)$ represents the external glucose infused in the bloodstream, whereas $U_I(t)$ is the exogenous insulin infusion; parameters are described in detail in [31]. By considering the two inputs, $U(t)$ and $U_I(t)$, and the two outputs, $x(t)$ and $y(t)$, the transfer function of a PID controller requires the tuning of the proportional, integral, and derivative gains. A possible methodology for their determination is to use an optimization algorithm, that tries to replicate the natural mechanism of insulin production in a healthy subject. In [39] different optimization algorithms are tested, genetic, particle swarm optimization, artificial bee colony algorithms. As an objective function it is chosen the normalized error, where the error is given by the difference between the value of the glucose in a healthy subject and the measured one

$$
Error = \frac{1}{n} \sum \left| \frac{x_{healthy} - x_{measured}}{x_{healthy}} \right|
$$

In Fig. 7, the block diagram of this procedure is shown.

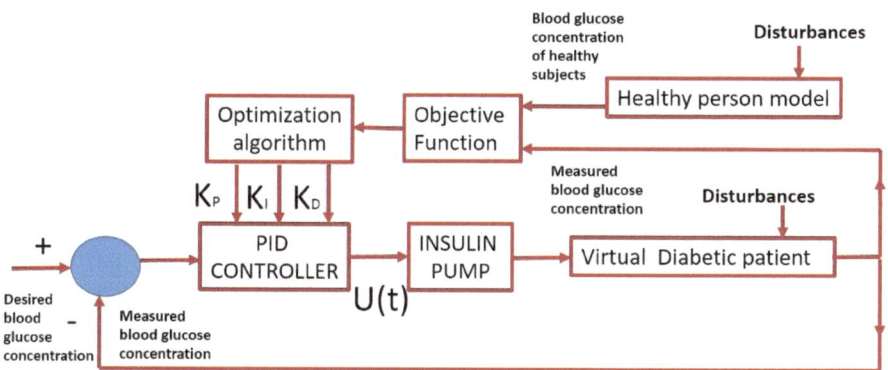

Fig. 7 Block diagram of the optimization procedure proposed by Soylu [39]

It could be appreciated the interdisciplinary characteristic of studying a physiological problem (in this case the diabetes control) with an engineering approach; to solve the problem with the proposed procedure it is required the model that simulates the "virtual diabetic patient" and the "healthy subject", the definition of the cost index, as function of the error, the choice of an optimization procedure, the knowledge of the physiological glucose concentration in a subject. For a real-case implementation also problems from sensors, and more in general, from a measurements points of view, must be faced to obtain the artificial pancreas that could be able to yield the right dosage of insulin, taking into account changes due to the meals, stress, physical activities, weather, and so on. Moreover, it could be noted the possibility of changing the adopted model with another one but preserving the general scheme of Fig. 7, as well as change the control strategy.

Epidemic Modeling and Control: the HIV/AIDS Spread

The Human Immunodeficiency Virus (HIV) is responsible of the Acquired Immune Deficiency Syndrome (AIDS); it infects cells of the immune system, destroying or impairing their function: the immune system becomes weaker, and the person is more susceptible to infections. The HIV/AIDS spread has been studied from a different point of view, at cells level [10, 20, 48], or by studying subjects' interaction, [18, 34].

AIDS is the most advanced stage of the HIV infection and can be reached in 10–15 years from the infection. This spread has some characteristics that must be taken into account in the modeling process:

- the HIV can be transmitted only by some body fluids;
- only about 54% of people with HIV are aware of the infection;
- currently, no vaccine exists, and the treatment consists of standard antiretroviral therapy to maximally suppress the HIV virus and stop its progression;
- using a condom and regular blood analysis on subjects belonging to risk-categories could help in contrasting the spread of this virus.

As far as the control aspect, the World Health Organization (WHO) suggests three levels of intervention: the first level of intervention is designed for healthy people to reduce the possibility of new infections and corresponds to the information effort to use wise attitudes; the second level of intervention is devoted to a fast identification of new infections to reduce the percentage of subjects that are not aware of their illness (and therefore to reduce new infections); the third level of intervention is the effort for medication to the aware infected subjects. Therefore, among infected subjects the most dangerous are those that, not aware of their status, could infect the unwise susceptible subjects. These considerations have recently suggested a new model describing the HIV/AIDS spread [19], in which two classes of susceptible individuals are introduced: $S_1(t)$ representing the number of healthy people that are not aware of unprotected sex acts risks (and then can contract the virus) and $S_2(t)$ denoting the number of healthy people that, suitably informed, gives great attention to the protection. Three classes of infected people are considered: the subjects $I(t)$ not aware of the infection, the patients $P(t)$ in the pre-AIDS condition and the subjects $A(t)$ in the AIDS state, see Fig. 8.

Fig. 8 Block diagram of the HIV/AIDS model proposed in Di Giamberardino et al. [19]

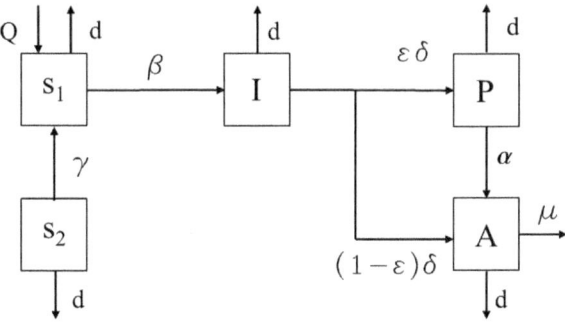

For the precise meaning of the parameters of the model see [19]; the complete controlled model, with the three levels of intervention, is

$$\dot{S}_1(t) = Q - dS_1(t) - \frac{\beta S_1(t)I(t)}{N_c(t)} + \gamma S_2(t) - S_1(t)u_1(t)$$

$$\dot{S}_2(t) = -(\gamma + d)S_2(t) + S_1(t)u_1(t)$$

$$\dot{I}(t) = \frac{\beta S_1(t)I(t)}{N_c(t)} - (d + \delta)I(t) - \psi \frac{I(t)}{N_c(t)}u_2(t)$$

$$\dot{P}(t) = \varepsilon\delta I(t) - (\alpha + d)P(t) + \phi\psi \frac{I(t)}{N_c(t)}u_2(t) + P(t)u_3(t)$$

$$\dot{A}(t) = (1 - \varepsilon)\delta I(t) + \alpha P(t) - (\mu + d)A(t) + (1 - \phi)\psi \frac{I(t)}{N_c(t)}u_2(t) - P(t)u_3(t)$$

After the usual model analysis with the determination of the equilibrium points (their number and nature depend on the value of $\beta - (d + \beta)$), the control strategy could be chosen in different ways. A possibility is to determine an optimal control by minimizing a suitable cost index, as in [21], where, taking into account resources limitation, the aim is to minimize the number of infected I(t) not aware of their condition; this number is not available in general, and a state estimator is required. The possible choice of a quadratic cost index suggested the use of a linear quadratic regulator and a linearization of the original system, thus yielding a solution in closed form. The results obtained consisted, as reasonable, in devoting a large amount of resource in inducing the people to check the eventual positiveness to the HIV, since the awareness of the infection has a double implication: it reduces dangerous contacts and induces patients to medication.

This example shows the use of control theory from a different point of view: analysis of the equilibrium points, linearization, state observer, linear quadratic control, convergence properties, just to mention the main items. In [10], the analysis of the equilibrium points is particularly interesting since it is shown how to drive the HIV patient state to the Long-Term Nonprogressor (LNTP) region of attraction.

A different approach [20] considers the model in [10], and introduces a switching control action that changes on the basis of the updated situation. The choice of a model and/or of a control strategy depends on the specific problem at hand, on the data available and of course on the goals to be pursued.

Bone Remodeling
Bone tissue during skeletal growth continuously adjusts its mass and architecture to changing conditions; the bones must adapt their shape and architecture efficiently to provide rigid levers for muscles, remaining as light as possible. At the cell scale, new bone tissue is formed by osteoblasts, and resorbed by osteoclasts; it is hypothesized that the osteocytes control bone adaptation, acting as mechanosensors based on local loading conditions. The understanding and possibly predicting the adaptation properties of bones are important for surgical screws, artificial joints, fractures, and so on. Bone remodeling formulation may be described from two different points of view: considering local regulation at the tissue and cellular level as in [13], or assuming global optimality of the bone structure, as in [1, 2, 24, 43]. As pointed out in Sect. 3, deviation from the remodeling equilibrium condition would initiate the remodeling activity, as a homeostatic goal.

In the second approach, the structure is usually divided into N elementary regions, the cellular automata CA; the driving force for adaptive activity is given by the difference between the strain energy averaged on the volume of the ith cellular automata, SED_i, and the strain energy density target value SED^*.

$$e_i(t) = SED_i - SED^*$$

In [1, 2] as remodeling local control rule it has been chosen a PID one, using the effective error:

$$\bar{e}_i(t) = \frac{1}{N+1}\left[e_i(t) + \sum_{j=1}^{n} e_j(t)\right]$$

$$\Delta x_i = x_i(t + \Delta t) - x_i(t) = c_P\bar{e}_i(t) + c_D[\bar{e}_i(t) - \bar{e}_i(t-1)] + c_I\sum_{\tau=1}^{t}\bar{e}_i(t-\tau)$$

being $x_i(t)$ the mass of the ith CA and c_P, c_I, c_D the PID parameters.

As said, the bones must adapt their shape and architecture efficiently remaining as light as possible; therefore, the goal to be pursued is to minimize the total mass, maximizing the stiffness, that is equivalent to minimize the total mass M and the total energy U.

$$J = (1 - \omega)\frac{M}{M_0} + \omega\frac{U}{U_0}, \quad 0 < \omega < 1$$

In [1, 2], the optimization problem has been solved also tuning optimally the PID parameters c_P, c_I, c_D, under box constraints. It has not been possible to solve the problem in closed form, and a numerical procedure based on sequential quadratic programming method is used. For the control modulus, it has been used a finite element analysis-based remodeling algorithm, able to implement the PID bone remodeling rule.

Also, in this case study, it can be appreciated the strong interdisciplinary required when facing this kind of problems; for this specific example, skills in structural engineering, biomedical materials, automatic control and optimization are required, as well as, of course, suitable knowledge of the physiological aspects involved.

Chemotherapy: An Optimal Control-Based Therapy

The determination of the chemotherapy, both in terms of drug dosage and scheduling, could be studied in the framework of optimal control theory, aiming at minimizing contrasting requirements, [33, 44]. Generally, the aim is to preserve healthy cells while destroying tumoral ones, using the minimum dosage of the drug to avoid toxic effects. In [23] it is considered the dynamics of the volume N of a tumor:

$$\frac{dN}{dt} = rNF(N) - G(N, t)$$

where F is the generalized growth function, r is the growth rate of the tumor; G describes the effects of the drug on the system; different choices are possible and *cell-kill strategies* are compared. As an example, it is now considered the Gompertzian growth for F and Skipper's log-kill hypothesis for G, as described in [23]:

$$\frac{dN}{dt} = rN \ln\left(\frac{1}{N}\right) - \delta u(t)N$$

being δ the magnitude of the dose, $u(t)$ is the control, that is the strength of the drug effect, assumed bounded $u(t) \in [0, M]$.

As for cost index a reasonable choice is to minimize the toxicity of the drug and the volume of the tumor at the end of the treatment, [23]:

$$J(u) = aN(T) + b \int_0^T u(t)dt, \quad a, b > 0$$

By introducing the change of variables $x(t) = \ln N$, the variation of the volume of the tumor may be rewritten as

$$\frac{dx}{dt} = -rx - \delta u(t)$$

The Pontryagin principle allows the determination of optimal control:

$$u^*(t) = \begin{cases} M & if \quad \lambda(t) > b/2 \\ 0 & if \quad \lambda(t) > b/2 \end{cases}$$

with $\lambda(t)$ the costate function satisfying the equation:

$$\frac{d\lambda(t)}{dt} = r\lambda(t), \quad \lambda(T) = ae^{x(T)}$$

The obtained control is the typical bang–bang solution, and the switching depends on the costate function that is, obviously, function of the volume of the tumor. Different choices of the cost index yield, obviously, different solutions; it is important the definition of the objective function that better considers the specificities of the drug and its toxicity. In the cost index, it could be introduced also the duration of the treatment as variable to minimize, and other elements such as the possibility of complications.

Pupil Light Reflex

The possibility of modeling pupillary light reflex and measuring the response of pupil also under different stimuli (auditory, for example) represents a support in the diagnosis of various pathologies. In particular, the latency after a stimulus is an indicator of drug and/or alcohol addiction, and of many diseases, such as the Parkinson, the Alzheimer, and the diabetes ones.

The interdisciplinary characteristic of physiological cybernetic is well evidenced when studying the pupillary light reflex, Fig. 9a. The determination of the pupil size implies the use of a pupillometer able to determine noninvasive measurements of the

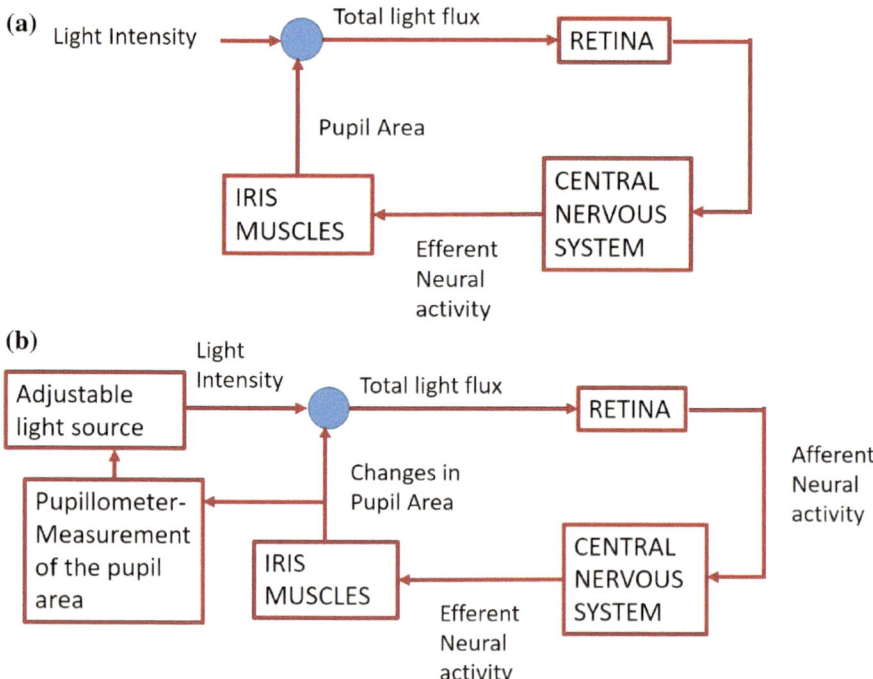

Fig. 9 Pupil light reflex; **a** feedback pupil light reflex scheme; **b** experiment proposed by Stark [42]

pupil and of its variations when there is a stimulus. This requires the implementation of methods of image analysis and processing, [16, 17, 28]; the images may be degraded by the natural fluctuations of the pupils, called *pupil noise*, present also in absence of any kind of stimuli, or by the presence of eyelashes, or artifacts due to the breath.

Stark [42] studied the fluctuation of the pupil by determining the transfer function of the pupil light reflex system and analyzed its stability with classical techniques, such as the Nyquist theorem and the Routh method. In Fig. 9b, it is described the experiment proposed by Stark to "open the loop", that is to deduce the transfer function of the pupillary reflex, [31, 42]. He introduced in the feedback loop a device with a light source able to adjust the delivered light at intensity inversely proportional to the pupil area. This expedient allowed him to deduce an estimation of the transfer function of the reflex.

The identification of a system by "opening the loop" is particularly important when studying physiological processes, since they operate generally in closed-loop; "open the loop" may imply a surgical action or the use of a pharmacological intervention, or, as in the case of pupil light reflex, the preparation of noninvasive experiments [31].

5   Conclusions

In this paper, it is discussed how physiological systems and their functions may be described within the framework of control theory; the feedback action allows the maintenance of the internal environments of the body, despite changes due to external conditions. Even if physiological phenomena may be adequately described by using control formalism, it is still not easy to find researches with skills in the disciplines involved and able to talk a common language.

Interdisciplinarity is the path for promising improvements to the understanding of physiological phenomena and to be able to act, whenever the spontaneous and natural intrinsic mechanisms do not work. This possibility is widely used in many fields with repercussions on everyday life, such as, for example, to determine the dosage of a drug.

The recent studies of gene networks make use of the classical and modern methodologies of automatic control: the formalism of system analysis and identification along with the feedback control theory have shown their power in modeling and analysis of gene regulatory network.

References

1. Andreaus U, Colloca M, Iacoviello D (2012) An optimal control procedure for bone adaptation under mechanical stimuli. Control Eng Pract 20(6):575–583
2. Andreaus U, Colloca M, Iacoviello D, Pignataro M (2011) Optimal-tuning PID control of adaptive materials for structural efficiency. Struct Multidiscip Optim 43(1):43–59
3. Beck CL (2015) Modeling and control of pharmacodynamics. Eur J Control 24:33–49
4. Behncke H (2000) Optimal control of deterministic epidemics. Optim Control Appl Methods 21:269–285
5. Campioni I, Notarangelo G, Andreaus U, Ventura A, Giacomozzi C (2012) Hipprostheses computational modeling: FEM simulations integrated with fatigue mechanics tests. Lect Notes Comput Vis Biomech 4:81–108
6. Cannon WB (1929) Organization for physiological homeostasis. Physiol Rev 9:399–431
7. Carley DW, Shannon DC (1988) A minimal mathematical model of human periodic breathing. J Appl Physiol 65:1400–1409
8. Carson E, Cobelli C (2014) Modeling methodology for physiology and medicine, Elsevier
9. Chaburn RL, Mireles Cabodevila E (2011) Closed-loop control of mechanical ventilation: description and classification of targeting schemes. Respir Care 56(1):85–102
10. Chang H, Astolfi A (2009) Control of HIV infection dynamics. IEEE Control Syst 28–39
11. Cobelli C, Dalla Man C, Sparacino G, Magni L, De Nicolao G, Kovatchev BP (2009) Diabetes: models, signals, and control. IEEE Rev Biomed Eng 2:54–96
12. Cosentino C, Bates D (2012) Feedback control in systems biology. Taylor & Francis
13. Cowin SC, Hegedus DH (1976) Bone remodeling I: theory of adaptive elasticity. J Elast 6:3
14. Das S, Caragea D, Welch SM, Hsu WH (2010) Computational methodologies in gene regulatory networks. Medical Information Science Reference
15. De Nicolao G, Magni L, Dalla Man C, Cobelli C (2011) Modeling and control of diabetes: towards the artificial pancreas. IFAC Proc Vol 44(1):7092–7101
16. De Santis A, Iacoviello D (2006) Optimal segmentation of pupillometric images for estimating pupil shape parameters. Comput Methods Programs Biomed 84:174–187
17. De Santis A, Iacoviello D (2009) Robust eye tracking for computer interface for disabled people. Comput Methods Programs Biomed 96(1):1–11
18. Di Giamberardino P, Iacoviello D (2017) Optimal control of SIR epidemic model with state dependent switching cost index. Biomed Signal Process Control 31:377–380
19. Di Giamberardino P, Compagnucci L, De Giorgi C, Iacoviello D (2018) Modeling the effects of prevention and early diagnosis on HIV/AIDS infection diffusion. IEEE Trans Syst, Man Cybern Syst
20. Di Giamberardino P, Iacoviello D (2018) HIV infection control: a constructive algorithm for a state-based switching control. Int J Control, Autom Syst 1–5
21. Di Giamberardino P, Iacoviello D (2018) LQ control design for the containment of the HIV/AIDS diffusion. Control Eng Pract 77:162–173
22. Doyle FJ, Huyett LM, Lee JB, Zisser HC, Dassau E (2014) Closed loop artificial pancreas systems: engineering the algorithms. Diabetes Care 37(5):1191–1197
23. Fister KR, Panetta JC (2003) Optimal control applied to competing chemotherapeutic cell-kill strategies. SIAM J Appl Math 63(6):1954–1971
24. Giorgio I, Andreaus U, Madeo A (2016) The influence of different loads on the remodeling process of a bone and bioresorbable material mixture with voids. Contin Mech Thermodyn 28(1–2):21–40
25. Glass L, Beuter A, Larocque D (1988) Time delays, Oscillations, and Chaos in physiological control systems. Math Biosci 90:111–125
26. Hovorka R, Canonico V, Chassin LJ, Haueter U, Massi-Benedetti M, Orsini Federici M, Pieber TR, Schaller HC, Schaupp L, Vering T, Willinska ME (2004) Nonlinear model predictive control of glucose concentration in subjects with type 1 diabetes. Physiol Meas 25:905–920
27. Huang C-N, Chung H-Y (2014) Applications of control theory in biomedical engineering. National Central University, Chungli Taiwan

28. Iacoviello D, Lucchetti M (2005) Parametric characterization of the form of the human pupil from blurred noisy images. Comput Methods Programs Biomed 77(1):39–48
29. Iacoviello D, Stasio N (2013) Optimal control for SIRC epidemic outbreak. Comput Methods Programs Biomed 110(3):333–342
30. Incremona GP, Messori M, Toffanin C, Cobelli C, Magni L (2018) Model predictive control with integral action for artificial pancreas. Control Eng Pract 77:86–94
31. Khoo MCK (2002) Physiological control systems. IEEE press series on biomedical engineering
32. Landi A, Laurini M, Piaggi P (2011) Physiological cybernetics: an old-novel approach for students in biomedical systems pp 48–62, Biomedical engineering–from theory to applications. www.intechopen.com
33. Ledzewicz U, Schättler H (2005) The influence of PK/PD on the structure of optimal control in cancer chemotherapy models. Math Biosci Eng 2(3):561–578
34. Naresh R, Tripathi A, Sharma D (2009) Modeling and analysis of the spread of AIDS epidemic with immigration of HIV infectives. Math Comput Model 49:880–892
35. Nowzari C, Preciado VM, Pappas GJ (2016) Analysis and control of epidemics. IEEE Control Syst Mag 36:26–46
36. Nugent ST, Finley JP (1987) Periodic breathing in infants: a model study. IEEE Trans Biomed Eng 34:482–485
37. Placidi G, Avola D, Iacoviello D, Cinque L (2013) Overall design and implementation of the virtual glove. Comput Biol Med 43(11):1927–1940
38. Placidi G, Avola D, Ferrari M, Iacoviello D, Petracca A, Quaresima V, Spezialetti M (2014) A low-cost real time virtual system for postural stability assessment at home. Comput Methods Programs Biomed 117(2):322–333
39. Soylu S, Danisman K (2016) Comparison of PID based control algorithms for daily blood glucose control. In: Proceedings of the 2nd world congress on electrical engineering and computer systems and science. pp 1–8
40. Stolwijk JE, Hardy JD (1974) Regulation and control in physiology. In: Medical physiology. pp 1343–1358
41. Swan GW (1984) Applications of optimal control theory in biomedicine. In: Dekker M (ed) New York
42. Stark L (1959) Stability, oscillations, and noise in the human pupil servomechanism. Proc IRE 47:1925–1939
43. Tovar A, Patel NM, Niebur GL, Sen M, Renaud JE (2006) Topology optimization using hybrid cellular automation method with local control rules. J Mech Des 128(6):1205–1216
44. Wang S, Schattler H (2016) Optimal control of a mathematical model for cancer chemotherapy under tumor heterogeneity. Math Biosci Eng 13(6):1223–1240
45. Warner A, Mittag J (2012) Thyroid hormone and the central control of homeostasis. J Mol Endocrinol 49:29–35
46. Wiener N (1948) Cybernetics, or control and communication in the animal and the machine. The MIT Press, Cambridge (MA)
47. Wu D, Zhang H, Cao J, Hayat T (2013) Stability and bifurcation analysis of a nonlinear discrete logistic model with delay. Discret Dyn Nat Soc 2013:1–7
48. Wodarz D, Nowak M (1999) Specific therapy regimes could lead to long-term immunological control of HIV. Proc Nat Acad Sci 96(25):14464–14469

New Computational Solution to Compute the Uptake Index from ^{99m}Tc-MDP Bone Scintigraphy Images

Vânia Araújo, Diogo Faria and João Manuel R. S. Tavares

Abstract The appearance of bone metastasis in patients with breast or prostate cancer makes the skeleton most affected by metastatic cancer. It is estimated that these two cancers lead in 80% of the cases to the appearance of bone metastasis, which is considered the main cause of death. ^{99m}Tc-methylene diphosphonate (^{99m}Tc-MDP) bone scintigraphy is the most commonly used radionuclide imaging technique for the detection and prognosis of bone carcinoma. With this work, it was intended to develop a new computational solution to extract from ^{99m}Tc-MDP bone scintigraphy images quantitative measurements of the affected regions in relation to the non-pathological regions. Hence, the uptake indexes computed from a new imaging exam are compared with the indexes computed from a previous exam of the same patient. Using active shape models, it is possible to segment the regions of the skeleton more prone to be affected by the bone carcinoma. On the other hand, the metastasis is segmented using the region-growing algorithm. Then, the uptake rate is calculated from the relation between the maximum intensity pixel of the metastatic region in relation to the maximum intensity pixel of the skeletal region where the metastasis was located. We evaluated the developed solution using scintigraphic images of 15 patients (7 females and 8 males) with bone carcinoma in two distinct time exams. The bone scans were obtained approximately 3 h after the injection of 740 MBq of ^{99m}Tc-MDP. The obtained indexes were compared against the evaluations in the clinical reports of the patients. It was possible to verify that the indexes obtained are according to the clinical evaluations of the 30 exams analyzed. However, there were

V. Araújo
Escola Superior de Biotecnologia, Universidade Católica Portuguesa, Porto, Portugal
e-mail: vania_dl_araujo@hotmail.com

D. Faria
Lenitudes – Medical Center & Research, Portugal, Universidade Católica Portuguesa, Porto, Portugal
e-mail: dborgesfaria@gmail.com

J. M. R. S. Tavares (✉)
Instituto de Ciência E Inovação Em Engenharia Mecânica E Engenharia Industrial, Departamento de Engenharia Mecânica, Faculdade de Engenharia, Universidade Do Porto, Porto, Portugal
e-mail: tavares@fe.up.pt

© Springer Nature Switzerland AG 2019

J. M. R. S. Tavares and P. R. Fernandes (eds.), *New Developments on Computational Methods and Imaging in Biomechanics and Biomedical Engineering*, Lecture Notes in Computational Vision and Biomechanics 33, https://doi.org/10.1007/978-3-030-23073-9_10

2 cases where the clinical evaluation was unclear as to the progression or regression of the disease, and when comparing the indexes, it is suggested the progression of the disease in one case and the regression in the other one. Based on the obtained results, it is possible to conclude that the computed indexes allow a quantitative analysis to evaluate the response to the prescribed therapy. Thus, the developed solution is promising to be used as a tool to help the technicians at the time of clinical evaluation.

Keywords Medical imaging · Image segmentation · Point distribution model · Active shape model · Bone metastasis

1   Introduction

The skeleton is most affected by metastatic cancer, with a higher prevalence for prostate and breast cancer. These two cancers cause in 80% of the cases, the appearance of bone metastases, which is considered the main cause of death. In many cases, the metastatic lesions are multifocal, which means that they are located throughout the skeleton with greater incidence in the axial skeleton [1].

The skeleton is constantly remodeling due to the coordinated activity of osteoclasts and osteoblasts. In normal bone, there is a balanced sequence: first, the osteoclasts absorb the bone and then the osteoblasts form bone in the same place. In cases of metastatic cancer, malignant cells secrete factors that affect this balance leading to osteoblastic stimulation [2].

Premature detection of metastases can prevent complications, control the stage of the disease, and help to determine the treatment to follow, which may result in a higher probability of survival and improvements in quality of life.

Bone scintigraphy with ^{99m}Tc methylene diphosphonate (MDP) is currently the most commonly used imaging technique in Nuclear Medicine to determine the extent of these lesions in the skeleton, as it provides a two-dimensional (2D) image of the skeleton showing regions with higher uptake (hotspots) [3]. In addition, it has good sensitivity and has been considered as the first alternative imaging method capable of diagnosing asymptomatic bone metastases, since it is readily available and provides a complete skeletal view at reasonable time and cost [4]. Modern bone scan techniques can detect an increase in bone mineral turnover as small as 10% in regions that are only a few millimeters in size. In contrast, a relatively large volume of bone (1 cm^3) must demineralize by about 50% before the change can be detected by radiographs. It is not surprising then, that in regard to prostate cancer, the bone scan is often used to stage patients and monitor the course of bone involvement. However, the interpretation of these exams has significant limitations: The evaluation of the exam is not yet standardized making the interpretations subjective and dependent on the experience of the technician. In numerous situations, these evaluations are described in vague terms as the presence or absence of tumor propagation in the skeleton. Therefore, a quantitative analysis of the images under study is necessary to reduce the variability of the observer in order to determine the extent of the lesions in the bone and to identify posttreatment changes that are clinically relevant.

To improve the monitoring of the treatment of bone lesions, some authors have developed scoring metrics for more objective methods of assessing the extent of bone metastasis, such as counting the number of lesions in the total skeleton, assessing the regional distribution of the metastasis or indexes that measures the tumor burden as a percentage of the total skeletal mass (Bone Scan Index) [5]. Therefore, the aim of the present study was to develop a semiautomatic method for the segmentation, i.e., identification, of regions of interest in ^{99m}Tc-MDP scintigraphy images of the skeletal system. The segmented regions allow the posterior assessment of the intensity of the hotspots under study and, therefore, the uptake index calculation.

2 Materials and Methods

2.1 Bone Scintigraphy

The bone scans used to evaluate the developed solution were obtained approximately 3 h after an intravenous (IV) injection of 740 MBq of ^{99m}Tc MDP. Whole-body images with anterior and posterior views were acquired according to a matrix size of 256×1024 pixels and using a gamma camera equipped with low-energy all-purpose collimators (Discovery NM 360, GE Healthcare). The energy discrimination was provided by a 20% window centered on the 140 keV of ^{99m}Tc.

2.2 Training Images Group

A training group of images was randomly chosen to build the Point Distribution Models [6, 7, 8, 9], used in the image segmentation step. The used group consisted of 10 images of patients who had undergone whole-body bone scintigraphy at Lenitudes Medical Center & Research, in Portugal.

2.3 Evaluation Images Group

The evaluation group consists of images acquired from 15 patients (8 males and 7 females), 7 of whom had prostate cancer, the other 7 have breast cancer, and 1 case of lung cancer. All these patients perform whole-body bone scintigraphy at Lenitudes Medical Center & Research every 3 months to evaluate the treatment response.

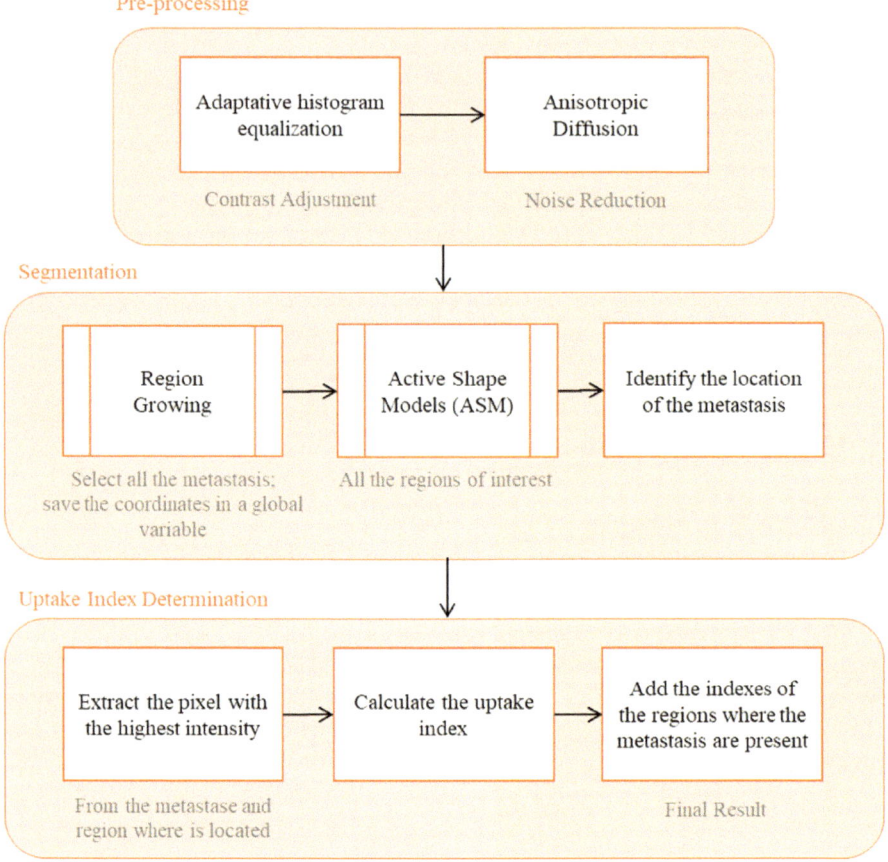

Fig. 1 Diagram of the proposed solution

2.4   *Bone Scintigraphy Image Processing*

The diagram of the developed solution is depicted in Fig. 1. The proposed solution has three main stages: image preprocessing, image segmentation, and uptake index computation. The image preprocessing stage is adapted to minimize the noisy arti-facts and enhance the contrast of the input bone scintigraphy images [10]. Image segmentation is one of the most common steps in image processing and analysis area, which intends to identify features of interested in input images [11, 12, 13]. Therefore, the enhanced images are submitted to the segmentation stage in order to identify the regions of the skeleton; namely, the skull, spine, thorax, clavicle, femur, humerus, pelvis, scapula, and sternum, Fig. 2.

After the segmentation of the regions under analysis, it is necessary to segment the existent metastasis, commonly known as hotspots. Then, each uptake index

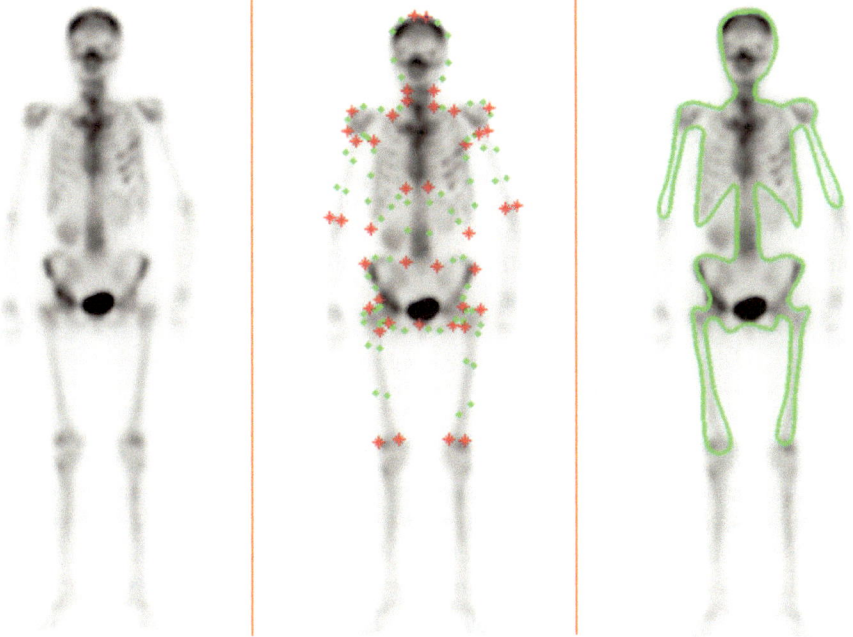

Fig. 2 Example of a point distribution model built to segment whole-body bone scintigraphy images: on the left, an example of a training image; On the center, landmark points used to build the model (117 points manually defined); on the right, automatically segmented whole-body scintigraphy image

assessment consists in computing the ratio between the value of the pixel with the highest intensity of the corresponding hotspot and the pixel with the highest intensity of the region where the hotspot is located without considering its region.

In the preprocessing step, an adaptive histogram equalization algorithm [14] is applied to enhance the contrast of the dark regions of the input images. In this step, it is also employed the anisotropic diffusion algorithm, first introduced by [15], which is a process that creates a space-scale system where an image leads to a parameterized family of images increasingly blurred based on a diffusion process [10]. This technique had become a useful tool to smooth image noise, detect image edges, segment images, and highlight them; particularly, anisotropic diffusion can smooth an input image while preserving the boundaries of the regions and the small structures present in the image [16].

As already mentioned, in the segmentation step, the Point Distribution Models (PDMs) proposed by Cootes and Taylor [6, 7, 8, 17] are used. PDMs have been used in statistical modeling of objects to describe, i.e., learning, their shapes from a set of training images. Thus, the built model describes the mean shape of the modeled object together with admissible variations in relation to the same mean shape [9]. PDMs emerged as a way of representing a set of forms of an object through the use of a flexible model of the position of certain landmarks located in image examples. These landmark points should reflect important characteristics of the shape of the object to be modeled, and must be selected in a similar way in all training images. In practice, this selection step is time-consuming and some automatic and semiautomatic methods have been proposed to define the landmark points to use in the PDMs building process [9].

In the process of building a PDM, the shape of the object to be modeled must be defined in a set of training images through a set of landmark points [8]. Once the points are selected, the coordinates of all n points that describe the i shape of the object are concatenated in vector x_i:

$$x_i = (x_{i1}, x_{i2}, x_{i3}, \ldots x_{in}, y_{i1}, y_{i2} \ldots, y_{in})^T \tag{1}$$

where $i = 1, \ldots, N$, with N representing the number of shapes in the set of training images and n the number of landmark points. Then, all the training shapes must be aligned in the same set of coordinates. After the alignment of the training shapes, it is possible to find the mean of the shapes and the variability presented in the training images. The modes of variation characterize the ways according to the landmarks of the built model tend to move, and can be obtained through a principal component analysis (PCA) to the derivations of the mean [9]. Thus, it is possible to rewrite each coordinates vector as

$$x = \bar{x} + P_s b_s \tag{2}$$

where x represents the number of points of the resultant shape of the modeled object, (x_k, y) is the position of landmark point k, $\bar{x}$ is the mean position of the landmark points, $P_s = (p_{s1} p_{s2} \ldots p_{st})$ is the matrix of the first t modes of variation, p_{si} correspond to the most significant eigenvectors in a PCA of the position variables, and $b_s = (b_{s1} b_{s2} \ldots b_{st})^T$ is a vector of weights for each variation mode of the shape. Each eigenvector describes how each landmark point moves on the training image set. Equation (2) represents the PDM of an object and can be used to generate new forms of the same.

Considering the existence of a trained model, i.e., a PDM, the corresponding Active Shape Model (ASM) [8] can be used to find, i.e., to segment, the modeled object in a new image. Starting with an approximate position of the object to be segmented, the ASM based segmentation applies an iterative optimization method to move each PDM landmark point to a better position. The decision-making to find the best position is based on finding the best combination of a local model along the normal boundary profile of the object in the image to be segmented [9].

Fig. 3 Example of the
regions used to determine the
uptake index

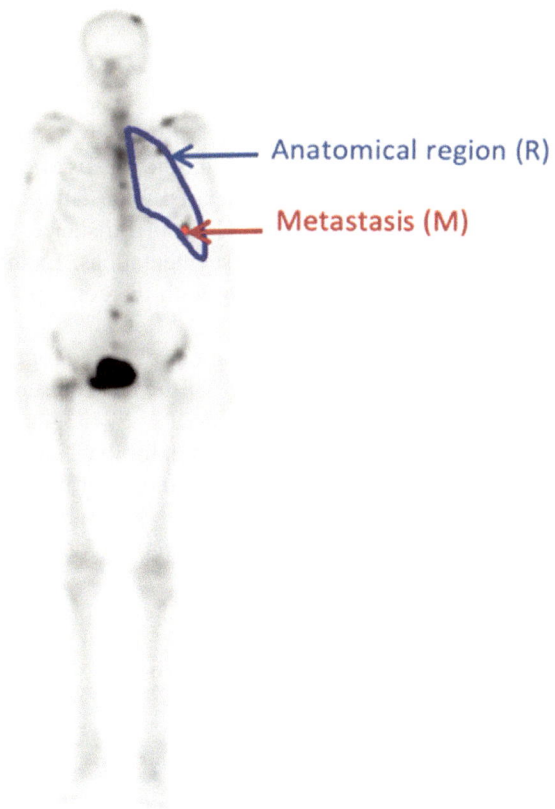

Anatomical region (R)

Metastasis (M)

2.5　Uptake Index Determination

In order to segment the metastasis, it was used the region-growing algorithm [12, 13], that allows the user to select a seed point and, from this seed point, i.e., seed pixel, the region to be segmented starts growing by attaching neighbor pixels that have similar properties.

The next step consists in calculating the uptake index based on the following steps (Fig. 3):

- For each segmented metastasis:

    a. Compute the tumor involvement based on the intensity of the image pixels, Fig. 3: in the anatomical region where the tumor is located, identify the pixel with the highest intensity of the region (R) that does not belong to the tumor region (M); identify the highest intensity pixel in the tumor region; calculate the ratio between the two intensities previously found;
    b. Add the computed result of the tumor involvement to the exam uptake index.

3   Results

3.1   Segmentation

The proposed computational solution was used on bone scintigraphy images, and each segmented region was compared against the corresponding manually segmented region. The Dice coefficient, Hausdorff distance, and centroid distance were used to validate the computational segmentations. Examples of segmentations obtained by the proposed solution along with the corresponding manual segmentations are shown in Fig. 4.

The computational segmentations obtained for the skull, thorax, pelvis, and thigh were then evaluated. The range of the centroid distance obtained for the skull was 4.49 ± 2.48, for the thorax was 4.78 ± 2.71, 3.44 ± 1.62 for the pelvis and 10.91 ± 6.59 for the thigh. On the other hand, for the Dice Coefficient, the obtained range for the skull was 0.89 ± 0.024, 0.89 ± 0.038 for the thorax, 0.91 ± 0.021 for the pelvis and 0.67 ± 0.073 for the thigh. Finally, the range of the Hausdorff distance obtained for the skull was 3.24 ± 0.53, 5.41 ± 0.56 for the thorax, 5.64 ± 0.72 for the pelvis and 3.98 ± 0.48 for the thigh.

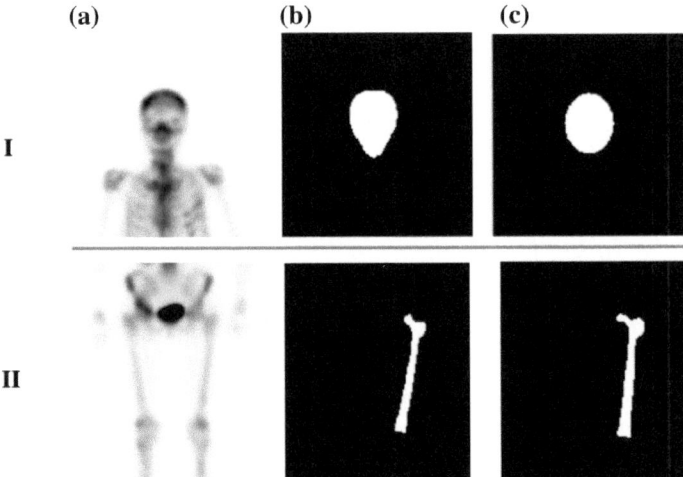

Fig. 4 Examples of segmentation obtained by the proposed solution for the skull (I) and of the femur (II): **a** training images; **b** segmentations obtained by the proposed solution; **c** segmentations manually delineated

Fig. 5 Uptake index values computed for the prostate cancer patients from their last two studies

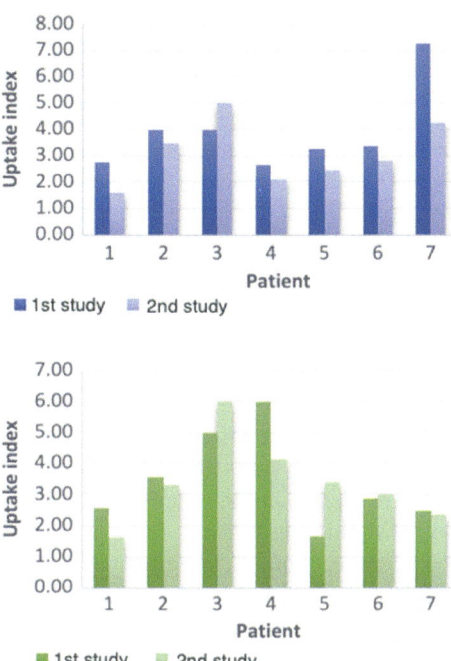

Fig. 6 Uptake index values computed for the breast cancer patients from their last two studies

3.2 *Uptake Index*

The uptake indexes were computed based on the approach previously described. For each patient, it was studied the last two exams, and the obtained indexes were compared with evaluations presented in the clinical reports.

As to the patients with metastatic prostate cancer, there was a decrease in the uptake index from the first to the second study in all patients with the exception of patient # 3, where the uptake index increased, as can be seen in Fig. 5. On the other hand, as to the patients of metastatic breast cancer, there was a decrease in the uptake index in the second study in patients # 1, # 2, # 4, and # 7 relatively to the first study. Contrary, in patients # 3, # 5, and # 6 the uptake index increased for the second study, Fig. 6. Finally, the patient with lung cancer in the second study had no metastasis and an uptake index of 1.98 computed in the first study.

4 Discussion

Regarding the distances between the centroids found for the skull, thorax, and pelvis, the mean of this metric ranged from 3 to 4 pixels, with a standard deviation around 2 pixels. Given that the size of the images under study was equal to 256×1024 pixels, the values found for this metric indicate high similarity between the computed and manual segmentations. Regarding the Dice coefficient, values close to 1 (one) show that the segmentations under comparison are similar, whereas closer to 0 (zero) means that there is no similarity between the segmentation. The mean of this metric found for the skull, thorax, and pelvis was 0.89 for the first two and 0.91 for the pelvis, which once again indicates that the segmentation under comparison were close. The Hausdorff distance, and the distance between centroids also shows that there is not a high degree of distinction between the segmentations. However, the comparison of the femoral segmentations showed more discrepant results. The distance between centroids was on average of 10 pixels with a standard deviation higher than the ones found in the previous cases, around 6 pixels. In terms of the Dice coefficient, the average of this metric was 0.67, considerably lower in comparison to the Dice coefficient found for the other cases. In fact, the segmentation of the femur obtained using the built Point Distribution Model generated the most distinct results. One way to solve this problem is to increase the number of training images. Another alternative would be increasing the number of landmark points distributed along the femur in the PDM building process; mainly, around the head of the femur, which is the less homogenous region to a segment.

In what concerns to the computed uptake indexes, the patients with prostate cancer had results that are in agreement with their qualitative evaluations. For example, in the cases where the study was described as "lower osteoblastic intensity", it was possible to verify a decrease in the uptake indexes, which indicates a good response to the prescribed therapy. However, in the case of patient #3, it was possible to verify a rise in the levels of uptake index, and the description of this study was evaluated as a "mixed response", this was due to the regression of some hotspots in the first examination and the appearance of new ones. However, when comparing the two studies through the uptake indexes, it was possible to verify the progression of the metastatic disease. To note, the case of patient #4 where the qualitative analysis was described as "overlapping hotspots in relation to the last study" and the indexes obtained showed a decrease from 2.65 to 2.1 suggesting improvements in the hypercaptation hotspots, mainly in the left ischium. In the case of metastatic breast cancer patients, all results are in agreement with the clinical evaluations. For the patient with carcinoma in the lung, it went from an index of 1.98 in the first study to an uptake index of 0 (zero), since there were no metastasis in the most recent study.

5   Conclusion

The challenges regarding the development of solutions for the fully automatic segmentation of the skeleton and metastasis in scintigraphy images remains a strong research topic. A semiautomatic solution for the segmentation of the regions of interest and the extraction of the information from these regions in ^{99m}Tc-MDP bone scintigraphy images was described. The developed solution proved to be effective in identifying the regions of interested in the input images. Although some difficulties have arisen in segmenting properly in some regions, these difficulties can be overcome by increasing the number of training images.

The developed solution was applied to 30 whole-body bone scans acquired from 15 patients. The computed uptake indexes were compared with the corresponding clinical evaluations, and a very promising matching was found. However, the proposed solution should be tested using more challenging cases in order to further evaluate and interpret critically the computed uptake indexes; mainly, to verify how they indicate properly the progression or regression of bone metastasis from ^{99m}Tc-MDP bone scintigraphy images.

References

1. Coleman RE (2001) Metastatic bone disease: clinical features, pathophysiology and treatment strategies. Cancer Treat Rev 27(3):165–176
2. Idota A, Sawaki M, Yoshimura A, Hattori M, Inaba Y, Oze I, Kikumori T, Kodera Y, Iwata H (2016) Bone scan index predicts skeletal-related events in patients with metastatic breast cancer. Springerplus 5(1):1095
3. Anand A, Morris MJ, Larson SM, Minarik D, Josefsson A, Helgstrand JT, Oturai PS, Edenbrandt L, Røder MA, Bjartell A (2016) Automated bone scan index as a quantitative imaging biomarker in metastatic castration-resistant prostate cancer patients being treated with enzalutamide. EJNMMI Res 6:23
4. Uemura K, Miyoshi Y, Kawahara T, Yoneyama S, Hattori Y, Teranishi J-I, Kondo K, Moriyama M, Takebayashi S, Yokomizo Y, Yao M, Uemura H, Noguchi K (2016) Prognostic value of a computer-aided diagnosis system involving bone scans among men treated with docetaxel for metastatic castration-resistant prostate cancer. BMC Cancer 16:109
5. Shimada H, Setoguchi T, Nakamura S, Yokouchi M, Ishidou Y, Tominaga H, Kawamura I, Nagano S, Komiya S (2015) Evaluation of prognostic scoring systems for bone metastases using single-center data. Mol Clin Oncol 3(6):1361–1370
6. Cootes T, Taylor C (2004) Statistical models of appearance for computer vision, University of Manchester
7. Cootes T, Hill A, Taylor C, Haslam J (1994) Use of active shape models for locating structures in medical images. Image Vis Comput 12(6):355–365
8. Cootes T, Taylor C, Cooper D, Graham J (1995) Active shape models-their training and application. Comput Vis Image Underst 61(1):38–59
9. Vasconcelos MJM, Tavares JMRS (2008) Methods to automatically build point distribution models for objects like hand palms and faces represented in images. Comput Model Eng Sci 36(3):213–241
10. Tsiotsios C, Petrou M (2013) On the choice of the parameters for anisotropic diffusion in image processing. Pattern Recognit 46(5):1369–1381
11. Jodas DS, Pereira AS, Tavares JMRS (2016) A review of computational methods applied for identification and quantification of atherosclerotic plaques in images. Expert Syst Appl 46:1–14
12. Ma Z, Tavares JMRS, Jorge RN, Mascarenhas T (2010) A review of algorithms for medical image segmentation and their applications to the female pelvic cavity. Comput Methods Biomech Biomed Eng 13(2):235–246
13. Oliveira RB, Filho ME, Ma Z, Papa JP, Pereira AS, Tavares JMRS (2016) Computational methods for the image segmentation of pigmented skin lesions: a Review. Comput Methods Programs Biomed 131:127–141
14. Pizer SM, Amburn EP, Austin JD, Cromartie R, Geselowitz A, Greer T, Romeny BTH, Zimmerman JB (1987) Adaptive histogram equalization and its variations. Comput Vis Graph Image Process 39(3):355–368

15. Perona P, Malik J (1990) Scale-space and edge detection using anisotropic diffusion. IEEE Trans Pattern Anal Mach Intell 12(7):629–639
16. Gulo CASJ, de Arruda HF, de Araujo AF, Sementille AC, Tavares JMRS (2016) Efficient parallelization on GPU of an image smoothing method based on a variational model. J R-Time Image Process. https://doi.org/10.1007/s11554-016-0623-x
17. Lindner C, Thiagarajah S, Wilkinson JM, Wallis GA, Cootes TF (2011) Short-term variability of proximal femur shape in anteroposterior pelvic radiographs. In: Proceedings of medical image understanding and analysis - MIUA 2011, London

Correction to: New Developments on Computational Methods and Imaging in Biomechanics and Biomedical Engineering

João Manuel R. S. Tavares and Paulo Rui Fernandes

Correction to:
J. M. R. S. Tavares and P. R. Fernandes (eds.),
New Developments on Computational Methods and Imaging
in Biomechanics and Biomedical Engineering, **Lecture Notes**
in Computational Vision and Biomechanics 33,
https://doi.org/10.1007/978-3-030-23073-9

In the original publication of this book, the volume number was incorrect. It should read "33" instead of "999". This has now been corrected.

The updated version of the book can be found at
https://doi.org/10.1007/978-3-030-23073-9